Studies in Big Data

Volume 95

Series Editor

Janusz Kacprzyk, Polish Academy of Sciences, Warsaw, Poland

The series "Studies in Big Data" (SBD) publishes new developments and advances in the various areas of Big Data- quickly and with a high quality. The intent is to cover the theory, research, development, and applications of Big Data, as embedded in the fields of engineering, computer science, physics, economics and life sciences. The books of the series refer to the analysis and understanding of large, complex, and/or distributed data sets generated from recent digital sources coming from sensors or other physical instruments as well as simulations, crowd sourcing, social networks or other internet transactions, such as emails or video click streams and other. The series contains monographs, lecture notes and edited volumes in Big Data spanning the areas of computational intelligence including neural networks, evolutionary computation, soft computing, fuzzy systems, as well as artificial intelligence, data mining, modern statistics and Operations research, as well as self-organizing systems. Of particular value to both the contributors and the readership are the short publication timeframe and the world-wide distribution, which enable both wide and rapid dissemination of research output.

The books of this series are reviewed in a single blind peer review process.

Indexed by SCOPUS, EI Compendex, SCIMAGO and zbMATH.

All books published in the series are submitted for consideration in Web of Science.

More information about this series at http://www.springer.com/series/11970

Ahmed A. Abd El-Latif · Bassem Abd-El-Atty ·
Salvador E. Venegas-Andraca ·
Wojciech Mazurczyk · Brij B. Gupta
Editors

Security and Privacy Preserving for IoT and 5G Networks

Techniques, Challenges, and New Directions

 Springer

Editors
Ahmed A. Abd El-Latif ⓘ
Department of Mathematics and Computer
Science
Faculty of Science, Menoufia University
Menoufia, Egypt

Salvador E. Venegas-Andraca ⓘ
School of Engineering and Sciences
Tecnologico de Monterrey
Monterrey, Mexico

Brij B. Gupta ⓘ
National Institute of Technology
Kurukshetra
Kurukshetra, India

Bassem Abd-El-Atty ⓘ
Department of Computer Science
Luxor University
Luxor, Egypt

Wojciech Mazurczyk ⓘ
Institute of Computer Science
Warsaw University of Technology
Warsaw, Poland

ISSN 2197-6503   ISSN 2197-6511   (electronic)
Studies in Big Data
ISBN 978-3-030-85430-0   ISBN 978-3-030-85428-7   (eBook)
https://doi.org/10.1007/978-3-030-85428-7

This Springer imprint is published by the registered company Springer Nature Switzerland AG
The registered company address is: Gewerbestrasse 11, 6330 Cham, Switzerland

Preface

Complexity is the worst enemy of security, and our systems are getting more complex all the time.

—Bruce Schneier, Data and Goliath: The Hidden Battles to Collect Your Data and Control Your World

The communication in 5G networks and beyond is currently in the center of attention of industry, academia, and government worldwide. 5G network concepts drive many new requirements for different network capabilities. As these networks aim at utilizing many promising network technologies, such as Software Defined Networking (SDN), Network Functions Virtualization (NFV), Information Centric Network (ICN), Network Slicing or Cloud Computing and supporting a huge number of connected devices integrating above mentioned advanced technologies and innovating new techniques will surely bring tremendous challenges for security, privacy and trust.

Based on the above mentioned technologies, it is predicted that communication in 5G networks will offer significantly greater data bandwidth and almost infinite capability of networking resulting in unfaltering user experiences for, among others: virtual/augmented reality, massive content streaming, telepresence, user-centric computing, crowded area services, smart personal networks, smart buildings/cities, and Internet of Things (IoT).

The concept of Internet of Things (IoT) incorporates various types of interconnected objects and devices which are typically resource-constrained but still have access to the Internet. The popularity of IoT has dramatically increased in recent years, as such solutions can be utilized for various purposes, including enhanced communication, transportation, education, business development, etc. However, in such networks still privacy and security aspects are among the most significant challenges which include lack of efficient and robust security protocols, lack of user awareness or improper configuration.

Considering above, secure network architectures, mechanisms, and protocols are required as the basis for 5G as well as IoT networks to address these issues and follow security-by-design approach. Moreover, since in such type of networks even more user data and network traffic will be transmitted, big data security solutions should be considered in order to address the magnitude of the data volume and ensure data security and privacy.

This book presents state-of-the-art research and the latest discoveries related to security and privacy-preserving mechanisms for 5G and IoT networks and their applications. Its chapters cover many important topics in this area, including traceability and tamper detection in IoT enabled waste management networks, secure Healthcare IoT Systems, data transfer accomplished by trustworthy nodes in cognitive radio, DDoS attack detection in Vehicular Ad-hoc Network (VANET) for 5G Networks, Mobile Edge-Cloud Computing, biometric authentication systems for IoT applications, and many others. Moreover, it proposes new models, practical solutions and technological advances related to transmitted data in 5G networks with IoT, cloud computing, and other applications in Smart Cities. Finally, it highlights and discusses the techniques and security challenges of data processing in IoT and 5G networks.

This book is intended to provide a relevant reference for students, researchers, engineers, and professionals working in this particular area or those interested in grasping its diverse facets and exploring the latest advances on security and privacy-preserving for IoT and 5G networks.

We would like to sincerely thank the authors of the contributing chapters as well as reviewers that for their valuable suggestions and feedback. The editors would like to thank Dr. Thomas Ditsinger (Springer, Editorial Director, Interdisciplinary Applied Sciences), Prof. Janusz Kacprzyk (Series Editor-in-Chief), and Ms. Rini Christy Xavier Rajasekaran (Springer Project Coordinator), for the editorial assistance and support to produce this important scientific work. Without this collective effort, this book would not have been possible to be completed.

We hope you will enjoy this book and this amazing research field of 5G and IoT networks security and privacy-preserving solutions!

Menoufia, Egypt   Ahmed A. Abd El- Latif
Luxor, Egypt   Bassem Abd-El-Atty
Monterrey, Mexico   Salvador E. Venegas-Andraca
Warsaw, Poland   Wojciech Mazurczyk
Kurukshetra, India   Brij B. Gupta

Contents

About the Editors

Dr. Ahmed A. Abd El-Latif received the B.Sc. degree with honor rank in Mathematics and Computer Science in 2005 and M.Sc. degree in Computer Science in 2010, all from Menoufia University, Egypt. He received his Ph.D. degree in Computer Science & Technology at Harbin Institute of Technology (H.I.T), Harbin, P. R. China in 2013. He is an Associate Professor of Computer Science at Menoufia University, Egypt and School of Information Technology and Computer Science, Nile University, Egypt. He is author and co-author of more than 140 papers, including refereed IEEE/ACM/Springer/Elsevier journals, conference papers, and book chapters. He received many awards, State Encouragement Award in Engineering Sciences 2016, Arab Republic of Egypt; the best Ph.D. student award from Harbin Institute of Technology, China 2013; Young scientific award, Menoufia University, Egypt 2014. He is a fellow at Academy of Scientific Research and Technology, Egypt. His areas of interests are multimedia content encryption, secure wireless communication, IoT, applied cryptanalysis, perceptual cryptography, secret media sharing, information hiding, biometrics, forensic analysis in digital images, and quantum information processing. Dr. Abd El-Latif has many collaborative scientific activities with international teams in different research projects. Furthermore, he has been reviewing papers for 120+ International Journals including IEEE Communications Magazine, IEEE Internet of Things journal, Information Sciences, IEEE Transactions on Network and Service Management, IEEE Transactions on Services Computing, Scientific reports Nature, Journal of Network and Computer Applications, Signal processing, Cryptologia, Journal of Network and Systems Management, Visual Communication and Image Representation, Neurocomputing, Future Generation Computer Systems, etc. Dr. Abd El-Latif is an Associate Editor of Journal of Cyber Security and Mobility, and IET Quantum Communication. Dr. Abd El-Latif is also leading many special issues in several SCI/EI journals.

Dr. Bassem Abd-El-Atty received B.S. degree in physics and computer science, M.Sc. degree in computer science, and Ph.D. degree in computer science all from Menoufia University, Egypt, in 2010, 2017, and 2020 respectively. He is currently an Assistant Professor in the Faculty of Computers and Information, Luxor University, Egypt. He is author and co-author of more than 30 papers, including refereed IEEE/Springer/Elsevier journals, conference papers, and book chapters. He is a reviewer in a set of reputable journals in Elsevier and Springer. His research interests include quantum information processing and image processing.

Prof. Salvador E. Venegas-Andraca received the M.Sc. degree in artificial intelligence and the D.Phil. degree in physics and the from the University of Oxford, in 2002 and 2006, respectively, and the M.B.A. (Hons.) and B.Sc. (Hons.) degrees in digital electronics and computer science from the Tecnológico de Monterrey. He is currently a Professor of computer science and the Head of the Quantum Information Processing Group, Tecnológico de Monterrey, Mexico. He is also a Leading Scientist in the field of quantum walks and cofounder of the field quantum image processing. His research interests include quantum algorithms as well as the algorithmic analysis of NP-hard/NP-complete problems. He has published more than 50 scientific articles, he has authored Quantum Walks for Computer Scientists, in 2008, the first book ever written on the scientific field of quantum walks, and coauthored Quantum Image Processing, in 2020, the first book totally focused on processing visual information using quantum systems. He has lectured in eleven countries across three continents and has been a Visiting Professor at Harvard University, the National Autonomous University of Mexico, del Valle University, Colombia, Bahia Blanca University, Argentina, and Yucatan University, Mexico. He is a fellow of the Mexican Academy of Sciences and a Senior Member of the Association for Computing Machinery.

Prof. Wojciech Mazurczyk (Senior Member, IEEE) received the B.Sc., M.Sc., Ph.D. (Hons.), and D.Sc. (habilitation) degrees in telecommunications from the Warsaw University of Technology (WUT), Warsaw, Poland, in 2003, 2004, 2009, and 2014, respectively. He is currently a Professor with the Institute of Computer Science, WUT, and the Head of the Computer Systems Security Group. He also works as a Researcher with the Parallelism and VLSI Group, Faculty of Mathematics and Computer Science, Fern Universitaet, Germany. His research interests include bio-inspired cybersecurity and networking, information hiding, and network security. He is involved in the technical program committee of many international conferences and also serves as a Reviewer for major international magazines and journals. Since 2016, he has been the Editor-in-Chief of the Open Access Journal of Cyber Security and Mobility. Since 2018, he has been serving as an Associate Editor for the IEEE Transactions on Information Forensics and Security and the Mobile Communications and Networks Series Editor for IEEE Communications Magazine.

Prof. Brij B. Gupta (SM'17) received the Ph.D. degree in information and cyber security from the Indian Institute of Technology Roorkee, India. He was a Visiting Researcher with Yamaguchi University, Japan, in 2015. He is currently guiding 10 students for their master's and Ph.D. research work in the area of information and cyber security. He has published over 90 research papers (including three books and 14 chapters) in international journals and conferences of high repute, including the IEEE, Elsevier, ACM, Springer, Wiley Inderscience, and so on. He has visited several countries, i.e., Canada, Japan, China, Malaysia, and Hong-Kong, to present his research work. His biography was selected and publishes in the 30th Edition of Marquis Who's Who in the World, 2012. He is also working principal investigator of various R&D projects. His research interest includes information security, cyber security, mobile security, cloud computing, Web security, intrusion detection, computer networks, and phishing. He has also served as a technical program committee member for over 20 International conferences worldwide. He is member of ACM, SIGCOMM, The Society of Digital Information and Wireless Communications, Internet Society, and the Institute of Nanotechnology, and a Life Member of the International Association of Engineers and the International Association of Computer Science and Information Technology. He is serving as an Associate Editor of the IEEE Access, an Associate Editor of IJICS, Inderscience and Executive Editor of IJITCA, Inderscience, respectively. He is also serving as a reviewer for the Journals of IEEE, Springer, Wiley, Taylor, and Francis. He is also serving as a guest editor for various reputed journals. He is also an editor of various international journals and magazines.

Authentic QR Codes for Traceability and Tamper Detection in IoT Enabled Waste Management Networks

H. Aparna, B. Bhumijaa, Ahemd A. Abd El-Latif, Rengarajan Amirtharajan, and Padmapriya Praveenkumar

Abstract With a population of around 136 crores, India is one of the world's largest developing nations. According to the Ministry of Housing and Urban Affairs, it is also one of the largest producers of solid waste, generating around 150,000 tons per day. The main problem lies with the management of the generated waste efficiency. Although the government has taken up several measures to ensure proper waste management techniques, most of them have not given the desired outputs. IoT (Internet of Things), being called one of the dominant infrastructure in the technological domain, has dynamic characteristics such as interoperability, scalability, self-configuring, and unique identity in the connected network. On the other hand, QR codes are machine-readable barcodes that contain an authenticated specific code that provides the tracking and routing information for the IoT enabled devices in the connected network. On applying the request-response communication model, the scanned QR data can be exchanged with the connected devices in the network. They can also be communicated with the centralised cloud server. Confidentiality, integrity, and Authenticity (CIA) can be integrated with the QR code using the Hash chaining principle.

Keywords QR codes · IoT · CIA · Tamper detection · Cloud platform

1 Introduction

In recent years, IoT has become the most prevailing and promising technology used worldwide. The logical setup of an IoT system and the importance of efficient communication within the network, request-response model, and various other communication models should be considered before designing an IoT network [1].

H. Aparna · B. Bhumijaa · R. Amirtharajan · P. Praveenkumar (✉)
School of Electrical & Electronics Engineering, SASTRA Deemed University, Thanjavur 613 401, India
e-mail: Padmapriya@ece.sastra.edu

A. A. Abd El-Latif
Menoufia University, Shibin El Kom, Egypt

The threats faced by various IoT systems and the need for threat detection mechanism have to be focused more on major communication, vulnerabilities and application layer protocols in IoT networks; threat mitigation mechanisms and how security services may get affected due to device limitation of devices should be concentrated along with the design [2, 3].

A simple but effective idea of using QR codes to mark and maintain college students' attendance has been proposed, and the information was made visible to the students to let them track. The performance of the students was collectively made available to the management too [3]. A QR code was designed using spatial and colour multiplexing of modules in this model. The QR thus obtained is proposed to have three times more storage capacity than a regular one, thus enabling efficient tracking when used in an IoT system, and the smart traceability was achieved through a mobile application [4].

Three layers of security for IoT applications are achieved using QR- code-based authentication mechanism (QRAM). QR codes are used for authentication purposes, and the user can gain access through scanning of QR alone and is proven to be highly robust and efficient [5]. The proposed method is a variant to the QR based detection mechanism, which is called Dual QR detection, which is proved to have a better bit error rate performance than regular conventional methods. The method is also shown to have less complexity in the K-best sphere detection model [6]. To enhance security and authentication, OTP and QR codes can be done collaboratively [7].

A method has been devised to prevent unauthorised users from reading sensitive information stored in the QR-code wherein the information is hidden and not visible to normal QR scanners. Also, a cover message is kept, shown on scanning with general scanners, thereby protecting the sensitive information. The confidential data is revealed only to the special scanner, which uses a key as an authentication mechanism. A mobile application that is to be used in a bus is designed here to enable the users to note the directions, route and payment option using QR code. Also, the application collects the users' valuable feedback at the end of their journey to test the app's suitability for the general public [8].

In most IoT networks, the communication is done using WiFi; a novel authentication method has been proposed, which uses QR code for the key exchange alongside Diffie-Hellman and SHA 256. The proposed method is proven to meet all the security requirements [9]. Pi, python and Open CV have been considered various platforms to recognise objects from the images taken by an IoT operated camera. The images and recognised data are stored in a web application [10].

QR codes and Dijkstra's algorithm are used to optimise the routes taken by a robot while moving; the QR codes are used as landmarks. Hence measurements are taken [11]. A mobile application has been created to make rail travel more efficient for the user. QR codes to know the vacant seats, booking food, and live tracking using GPS are the main features of the proposed application [12].

The proposal for using QR codes in power telecommunication is put forth, where a power telecommunication tag is designed using the QR codes [13]. The proposed method uses a mobile application to track and monitor patients in a health care institution. It is a framework, QR codes are used to track and monitor patients undergoing rehabilitation. The results proved to be highly effective [14]. QR code-based mutual authentication in IoT network can be used for key exchange and is proven to be highly robust [15].

Multi-layered hashing has been proposed, which is called "Hash Vine". The method is proven to have a negligible number of collision and is efficient and robust. Hash vines are shown to have produced long hashing chains which can be used for multiple authentication purposes. Two different hash functions are used in the proposed method to produce an infinite length hash chain [16]. Request–response or request–reply is one of the basic methods computers use to communicate with other devices connected in a network, in which the first computer sends a request for data and the second responds to the request. More specifically, it is a message exchange pattern. A requester sends a request message to a replier system, which receives and processes the request, ultimately returning a message in response [17].

Due to dual hash functions, secrecy is maintained here perfectly and is forwarded efficiently [18]. Using the binomial tree model, a methodology has been developed to improve the time and space complexity of reverse hash chain traversal. The idea proposed is proven to reduce complexity and be applied for chains of infinite length too [19]. 2D-backward hash key chains have been used to obtain a "continuous secure scheme". Compared to other methods and a 1D hash key chain, this method is very effective in terms of randomness and assuring integrity [20].

An efficient method has been proposed, which uses the hash chaining principle to obtain a multi-layered authentication mechanism. The proposed method depends on the receiver and can tolerate packet loss [24]. A seed value is used to generate a hash chain of a particular length. The chain is self-updating in nature using the erasure coding algorithm, and after the length has been achieved, the seed value is taken from the last value to obtain a new chain [25]. A self-renewing hash chain method is proposed where the one-time signature is used in combination. Linear Partition combination algorithm is used to generate a self-updating hash chain [26]. To secure RFID tag, hash chain and 3-way handshake security methods were used as a combination to ensure integrity and authenticity [21–23, 27–34]. Also, end-to-end security is very important to ensure the confidentiality and secure acceptability of the transimitted data [35–39]. A comparison has been made with the recently available work in literature with the proposed scheme (Table 1).

On analysing the various procedures and methodologies on IoT, QR codes, hash and request-response models, this chapter highlights the following points:

- The proposed system integrates IOT with nested QR codes for easy and hassle-free data transfer.
- QR codes are fast, secure, and the safest means for accessing data.
- To attain dual storage capability without having to increase the area, Nested QR codes are used.
- The QR codes are protected using the hash chaining principle against tampering and eavesdroppers.

Table 1 Comparison of the proposed scheme with the available literature

Ref.no	Methodology in literature	Proposed methodology
Bakar et al. [4]	This paper makes use of normal QR codes for tracking and monitoring student attendance	The proposed methodology uses Nested QR and checks tampering the QR code, thereby ensuring a more secure interaction with the use
Ramalho et al. [5]	In this model, a QR code with triple storage capacity as the normal QR is developed to track banknotes. However, the printing process for this unique QR is complex, although the cost is low	In the method proposed, the nested QR has a storage capacity twice as that of the normal QR within the same area, and the cost and complexity of printing is very low
Lin and Chen [21]	This model proposes a QR code that has a hidden message which can be scanned using a special scanner which is available only to authorised users	A normal scanner is sufficient to scan the Nested QR, and the authentication is also achieved in the proposed method without any special external resources
Arebey et al. [22]	This method makes use of RFID and a Camera for monitoring waste collection and management, which may be expensive, and the cost of replacement in case of damage will be even more additive to the initial	The proposed method uses QR codes that can be easily created and printed onto any surface and, in case damaged, can easily be replaced at almost zero additive cost
Krithika et al. [23]	This paper makes use of Raspberry Pi along with other sensors for monitoring waste in bins. For this method to function efficiently, there is a need for a constant source of power supply	The proposed method doesn't make use of any sensors or microcontroller unit. Yet tracking of waste is efficiently performed at a very less cost and without any requirement for a power source

- Further, the system makes use of Request-Response model, which helps the legitimate users connected over a network to access the data in very minimal time.
- Interconnection was done through a user-friendly mobile application that helps track, intimating the users through alert message.
- Storing the data in the cloud servers enables more security and provides quick means of data transfer.

2   Preliminaries

2.1   Internet of Things (IoT)

It describes a network of interconnected things or devices connected via the internet or LAN through the medium of which the objects can transfer data with each other and work in coordination with each other. IoT also allows things to be operated from afar through the internet. It allows the remote control of the operations of these devices along with automation in some cases. Although there are a few concerns concerning safety and data security, IoT is widely being employed worldwide. IoT generally comprises three kinds of devices: Devices that can send collected data to other devices, Devices that receive data store it, and perform actions based on the conclusions drawn from it and devices that can simultaneously do both the above-mentioned activities. Sensors are devices that are generally used to send and receive information. Sensors, for example, light sensor, heat sensor, moisture sensor etc., are used to collect information from our surroundings and send it to a receiver or a microcontroller unit for further process.

To act on the information collected, usually microcontroller units or devices comprising such units are used. These devices smartly draw conclusions based on the information collected and sent by the sensor and perform the necessary actions. A few devices can perform both reading and sense of data collected; such devices are generally called systems as they are a group of small devices and sensors. These systems perform tasks which are necessary by analysing the surrounding through the data collected by the sensors. The analysis is performed through a processor or a controller unit built within the system. The devices within such systems are connected using LAN or via the internet too. When the devices are connected to the internet, the data being transferred or collected can easily be stored in the cloud server rather than in the case of LAN. Also, data accessibility is wide in internet-connected devices as they can be remotely controlled from anywhere around the world without hassle.

2.2   Quick Response Code (QR)

It is a two-dimensional bar code that is secure and can be easily created to store data. It consists of black squares or boxes printed on the white background. The QR code can be printed on to any surface easily can be read without any hassle using a simple barcode scanner or smartphone camera. It was initially created to keep track of cars in a manufacturing unit. But today, QR codes are being used for various other purposes too, including tracking and monitoring. Nested QR is one variant of QR codes, wherein two separate QRs are embedded over each other. The outer QR is larger in size when compared to the inner one. The inner QR is rotated by 90, 180 or 270° and placed over the outer QR, as shown in Fig. 1. Both the QR codes can be read by the normal scanner itself, except that the angle and distance have to be changed for reading individual codes.

2.3   The Nested QR Code

Quick Response (QR) code is a two-dimensional barcode that can be easily printed on any surface. It is safe as once generated, the content of the code cannot be changed

Fig. 1 **a** Quick response
code. **b** Nested quick
response code

(a) **(b)**

unless a new code is generated. In the proposed method, a Nested QR code is initially
generated based on an Identity number unique for each bin and a mobile number
registered with the waste managing company.

The Nested QR code comprises two independent QR codes, generally referred to
as the Outer QR and the Inner QR codes. The Outer QR code comprises the bin's
identity number and its tampering check verification code in the method proposed
here. The Inner QR is the user mobile number registered with the trash collection
agency or the municipal corporation. Nested QR is formed by embedding the Inner
QR on to the Outer QR after rotating it by either 90,180 or 270 degrees. The Outer
QR is almost three to five times the size of the Inner QR, and the Inner QR is placed
at almost the centre of the Outer QR, a position which does not affect the readability
of both the QR codes. The Nested QR can be read using a normal barcode scanner
or QR reader.

For reading both the QR codes, we need to change the angle and distance from
which each QR code is read. The Nested QR codes have more storage capacity when
compared to regular QR codes, as they can store two sets of individual messages
within the same area, where regular QR just stores one. Also, this kind of QR code
is more robust and safe from attacks when compared to regular ones, although the
regular ones are safer too. The Nested QR has all the general properties of QR with
extra features, making it a novelty for the proposed methodology. This QR code, thus
generated, is stuck against the bin in the customer's household.

2.4 Cloud Storage and Cloud Database

Storing data over the internet in a space offered by some agency is known as cloud
storage. Cloud storage allows tons of data to be stored online and relieves the user
from storage limitations. Also, cloud storage offers security, and only authenticated
personnel have access to the data. Cloud allows multiple user access along with variable levels of permissions. This ensures that only properly authenticated users can
access reading the data and make changes to it if required. It is highly synchronous
in nature, as the changes made in data is immediately reflected. If an internet facility

is available, the cloud can be accessed from anywhere, thereby making the transportation of data using flash drives or pen drives almost unnecessary. In case of any disaster, the data can be easily recovered without any chaos. It is a very convenient mode to transfer large files and is also cost-efficient in nature.

2.5 MIT App Inventor

The MIT App Inventor is a very useful tool for beginners to learn designing Android-based mobile applications. Massachusetts Institute of Technology has designed it. The app inventor has several components which can be integrated according to the requirement of the app being designed. It allows the designing both front end and back end by integrating "blocks" each of which has different functionalities.

2.6 Smart Waste Management (SWM)

It is just not managing existing waste, but it efficiently's the management of waste along with the utilisation of technology. It also includes the aspect of reducing the amount of waste being generated. Proper disposal techniques and constant tracking and monitoring of the waste existing is an important aspect involved here. Smart waste management also encourages the reusing and recycling of waste as much as possible.

2.7 Request Response Model

Every client is allowed to access information by scanning the QR code eliminating the necessity to log into a terminal mode. Servers have better control access and resources to ensure that only authorised clients can access the data and server updates are administered effectively as shown in Fig. 2.

Applications used for this model is built regardless of the hardware platform or technical background of the entitled software providing an open computing environment, enforcing users to obtain the services of clients and servers. Advanced database products enable user/application to gain a merged view of data dispersed over several platforms. Rather than a single target platform, this ensures database integrity to perform updates on multiple locations enforcing quality recital and recovery. This model has better efficiency, ensures reliable delivery, and the process consumes almost no time. By applying the request-response communication model, the scanned QR data can be exchanged with the connected devices in the network and communicated with the centralised cloud server.

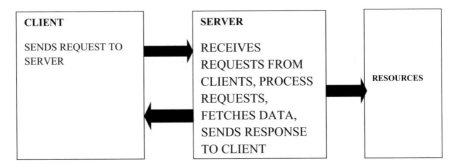

Fig. 2 Block representation of request response model

2.8 Hash Chaining

As the communication between the scanner and the QR code is unprotected in some cases, eavesdroppers can listen in the immediate vicinity. With a basic system, the integrity of the data contained in the QR is not guaranteed, as the data contained in the code can be manipulated. Using database servers to store information about the code with lookup tables helps in avoiding these issues. Using secure distribution channels for readers and mutual authentication between client and server can deter such attacks. Unauthorised readers or readers which have been tampered with must not be able to authenticate valid QR falsely. Furthermore, a hash chain or general hash function acts as a provision against random failures. Hashing is a technique used to uniquely identify a specific object from a group of similar objects, as shown in Fig. 3.

A hash function can be used to map a data set of arbitrary size to a data set of a fixed size, which falls into the hash table. Hash chains involve the repeated application of a cryptographic hash function to a message to securely authenticate a user a finite number of times using a one-way encryption function. When 2 keys are hashed or pointing to the same index number, a collision is set to occur. Collision may lead to undesired results and hence need to be avoided. To mitigate collision, chaining principle is made use of, wherein each index number points to an independent linked

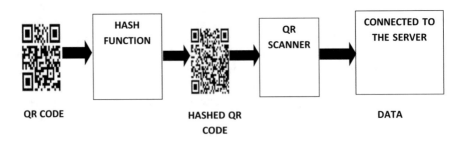

Fig. 3 Hash chaining principle

list, thereby allowing multiple keys to point to the same index. All these keys are stored in the linked list. Also, chaining principle allows insertion in constant time, and the table can grow infinitely long till space is available.

3 Proposed Methodology

The Quick Response codes are generally safe and secure and very difficult to tamper, making it preferable. To increase the robustness as well as the storage capacity, a new form of QR called the "Nested QR" is being used here, wherein one QR code is embedded over the other as shown in Fig. 4. This is done to attain dual storage capability without having to increase the area. The QR code is passed through a hash function to obtain a hashed QR. Generally, hashing is used to identify an element from a list of similar elements uniquely. Sometimes collision can take place when two keys point to the same index. In order to avoid that, a method called chaining is used wherein each index of the array is a linked list independently. This makes the insertion happen in a constant time. If there is enough space, the hash chain can grow infinitely long. On scanning the QR code attached to IoT devices, it can directly open up a website or a cloud server or a database where the complete information about the device can be stored or open up after verifying the integrity and authenticity of the scanning personnel. Tamper detection, read/write/modify access on the QR code can be controlled and monitored using cloud-enabled platform.

To trace and tamper detection, a mobile application has been designed. The design has been made using MIT App Inventor. The application has multiple components attached to it, through which the tracing is achieved. An inbuilt QR/Barcode scanner is present, which scans the code and reads the details. Once the details are read, the location and time sensor or clock components of the application notes down the location date and time of scanning, enabling the tracing.

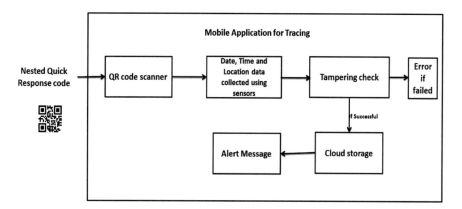

Fig. 4 Block diagram of the proposed methodology

Hence, the details can now be stored in the cloud database, which has been integrated with the application. But before storing it in a database, a tampering check is done, which is almost similar to checksum. In case there is a mismatch between expected and obtained data, then an error message is shown. If no error has been found, then the application proceeds to store the data in the cloud. After the storage process has been completed, an alert message is sent to the mobile number linked to the application to notify the scanned QR code. This will ensure that the owner has immediate information that a scan has been performed even before accessing the cloud data.

Upon scanning the QR code, the check for any kind of tampering of QR is done [40]. If the data present in QR passes the test, then date, time, and location where the trash was collected is updated in the cloud storage. If the tampering test fails, then an error message is shown on the screen, prompting the user to contact the authorities. Once the details are stored in the cloud database, the customer receives an intimation regarding the status of trash collection through an alert message to the registered mobile number. This enables both the customer and the trash managing company to keep track of the collection process.

4   Results and Discussion

Waste management is very much essential to keep balance in nature. Improper waste management will lead to bio and ecological imbalance. Living organisms in the surroundings will get affected by the accumulation of waste, leading to severe and deadly diseases. Dumping of waste also leads to the contamination of land and water bodies resulting in depletion of natural resources. Sometimes harmful chemicals generated from the waste may seep into the land and affect the groundwater. Efficient management of waste can easily reduce land and water pollution and control the spread of diseases. Several methods have been proposed for waste management using IoT, which can resolve waste management problems on proper implementation. The proposed method is easy to implement on a large scale as QR codes can easily be printed and circulated, and also smartphones are almost available to everyone which will make it easy for the application designed to be used. Further, the application can be developed in several ways to improve its utility, and also it can be used very efficiently in the future.

A mobile application is created for monitoring the trash collected from the households. The MIT app inventor has been used for this purpose. The app inventor was originally designed by Google in collaboration with the Massachusetts Institute of Technology to enable learners worldwide to easily learn app designing and development and is an open source and free software that can be used over the internet. The inventor uses JAVA, KAWA and Scheme as its background languages over which all the codes for the functions are written. Similar to languages such as Scratch, the app inventor makes use of Graphical User Interface or GUI. Using this GUI, the users can choose the design and components required for the application they intend to

Fig. 5 Opening screen of
the AI companion App

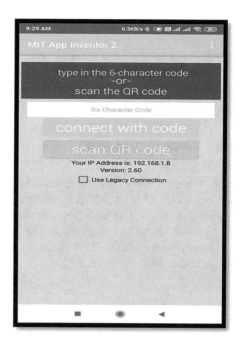

design. It has a drag and drop kind of mechanism that allows the learner to pick and attach different blocks together and make different applications. Each of these blocks has different functionality by itself and, when attached, work in sync to generate the required application. The app inventor has collaborated with the Google Firebase cloud server, which allows cloud-based storage applications to be enabled via the applications being designed by learners.

To run these apps designed on an android based smart phone, there is an AI companion app that allows the designed application to run on the android device without having to install the APK of the app. The AI companion app is linked to the app inventor via a code. Until the companion app is connected, the application designed in the app inventor can be tested on the android device, as shown in Fig. 5.

The application designed here consists of a QR scanner, tampering check, cloud storage and a customer alert feature. After designing the display screens of the app, various sensors and modules required for the implementation of features are added and enabled via the app.

4.1 Barcode Scanner

A Barcode Scanner is used for scanning the QR. Both the QRs in the nested QR model can be scanned using the same scanner. In the app inventor, designing a hidden component called the barcode scanner is added to the app, which prompts the system to scan the QR code whenever called. Usually, the barcode component collaborates with the android device's inbuilt QR scanner to read the code. But in case there is no inbuilt scanner present, then a scanner app needs to be installed. The companion app works in coordination with this scanner when permissions are enabled and scans the code whenever called for.

4.2 Clock Component

A clock component is added to the mobile application, which takes the system's date and time. This is also a hidden component and is not displayed on the screen of the application design. The clock has many inbuilt features. It can display date, day, and time in both 12 and 24 h. Format. It also can easily be integrated with other features and can give accurate results. The clock component also allows time in any format, which allows us to show the time in any time zone, GMT or IST.

4.3 Location Sensor Component

The location sensor component is also integrated, which senses the location using the GPS of the mobile device, giving the latitude and longitude of the scan place. The location sensor gives the latitude and longitude and gives the feature of giving the location's address. The accuracy of the location can also be changed, and also one can set the latitude and longitude of a particular location. It can also measure the change in location within a time interval. It can measure the altitude too.

4.4 Tampering Check Mechanism

Tampering check is performed in a format similar to that of checksum verification on the Identity number of the bin stored in the Outer QR code of the Nested QR. The Identity number is first taken as 2 digits at a time, representing 8 digits in binary and are added together to obtain the final sum, which is again an 8 digit binary number. This number is subtracted from 255, which is the last 8 digit binary number, to obtain the checksum. This checksum is converted to Hexadecimal and is compared with the value scanned from the QR. If there is a match, then the tampering check is said to be successful, and the application proceeds with storing the value in the cloud database. Else if there is a mismatch of values, then the tampering check is said to have failed, and an error message is shown on the screen asking the user to contact the authorities regarding the same. Since this method follows the same procedure as that of the normal checksum mechanism, the tampering check mechanism is robust and immune to attacks.

4.5 Storing of Data

Storage is achieved through a cloud server database. Cloud enables storage of a large amount of data that can be retrieved easily on having proper permission. It offers data security which is very much required, along with easy accessibility. Cloud also allows multiple users to access the database It is very convenient as it is accessible from anywhere through the internet. Google Firebase database was used for cloud storage here. Firebase is a Google server-based platform, employing the "platform as a service" formula, which has several features: real time database, ability to design web applications, and mobile applications. All these features are available in the cloud and hence easy to access and arc safe to work. The firebase provides all the back end mechanisms that are generally difficult for inexperienced workers, making it easy for all types of users. It can work both with android and ios platforms.

The real-time database feature of the Firebase was integrated with the app inventor using an URL generated by the database server. It is enabled to store real-time data. The real-time database allows the user to define security rules by themselves and when required with proper authentication and permissions. This allows the user to decide the level of security required for the data stored in the cloud. The database also issues warning message in case the security features decided by the user are weak thereby making the user aware of the various problems which may come in the future.

4.6 Alert Message Feature

For the customer alert feature, messaging feature of the app is used. Outer code with ID, inner QR code with the Mobile number and the nested QR are shown in Fig. 6. The messaging feature collaborates with the SMS sending mechanism of the

| (a) | (b) | (c) |

Fig. 6 **a** Outer QR code, with the ID number. **b** Inner QR code with mobile number. **c** The Nested QR Code

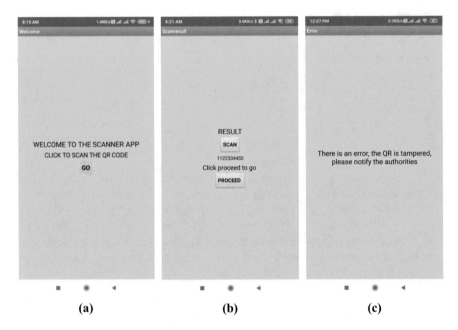

Fig. 7 **a** Welcome screen of the application. **b** On scanning the code, the ID gets scanned and is shown on the screen. **c** Error screen after tampering check is failed

android mobile device to send an alert message. The app sends a pre-written and stored message to the registered mobile number as soon as the feature is enabled (Fig. 7, 8).

5   Conclusion

An IOT based network is built for efficient means of collecting trash from households and for hassle-free monitoring and tracking of the waste collected. The proposed method uses Quick Response (QR) codes to track and monitor the waste collection procedure. The QR codes are designed to be scanned via an android application, which verifies, stores data and alerts the user. QR codes are used as they are safe and can be printed on any surface, and scanned easily using smartphones. The proposed method uses an improved version of a normal QR code called the Nested QR, which can store more data without compromising the area. Further, data stored in the QR is secured and is protected against tampering. Also, scanning of QR code immediately alerts the user and updates the time and place of disposal in the cloud database, making it easy to monitor and track the trash collected. Thus, through this project, an efficient and robust means of trash collection is designed using recent technologies, creating an easy and convenient means of trash collection and monitoring.

(a)

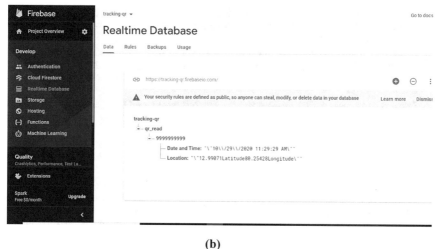

(b)

Fig. 8 **a** Alert message being sent. **b** The information-date, time and location being stored in the cloud database

Acknowledgements The authors wish to acknowledge the Science and Engineering Research Board (SERB), Department of Science & Technology (DST), India (EEQ/2019/000565), for providing the financial support to carry out this research work. The authors also wish to acknowledge SASTRA Deemed University, Thanjavur, India, for the infrastructural support.

References

1. Bahga, A., Madisetti V.: Internet of Things: a Hands-on Approach (2014)
2. Nebbione, G., Calzarossa, M.C.: Security of IoT application layer protocols: challenges and findings. Future Internet **12**, 55 (2020)

3. Belghazi, Z., Benamar, N., Addaim, A., Kerrache, C.A.: Secure wifi-direct using key exchange for IoT device-to-device communications in a smart environment. Future Internet **11**(12), art. no. 251 (2019)
4. Bakar, SA, Salleh, S.N.M., Rasidi, A., Tasmin, R., Hamid, N.A.A., Nda, R.M., Rusuli, M.S.C.: Integrating QR code-based approach to University e-class system for managing student attendance. In: Advances in Intelligent Systems and Computing, vol. 1158, pp. 379–387 (2021)
5. Ramalho, J.F.C.B., Correia, S.F.H., Fu, L., Dias, L.M.S., Adão, P., Mateus, P., Ferreira, R.A.S., André, P.S.: Super modules-based active QR codes for smart trackability and IoT: a responsive-banknotes case study. NPJ Flexible Electron. **4**(1), art. no. 11 (2020)
6. Al-Ghaili, A.M., Kasim, H., Othman, M., Hashim, W.: QR code based authentication method for IoT applications using three security layers. Telkomnika (Telecommunication Computing Electronics and Control) **18**(4), 2004–2011 (2020)
7. Syed Moinuddin Bokhari, B., Bhagyaveni, M.A.: Dual QR decomposition in K-best sphere detection for internet of things networks. Cluster Comput. **22**, 7713–7722 (2019)
8. Zhao, Q., Yang, S., Zheng, D., Qin, B.: A QR code secret hiding scheme against contrast analysis attack for the internet of things. Secur. Commun. Netw. art. no. 8105787 (2019)
9. Fong, S.L., Wui Yung, D.C., Ahmed, F.Y.H., Jamal, A. Smart city bus application with quick response (QR) code payment. In: ACM International Conference Proceeding Series, Part F147956, pp. 248–252 (2019)
10. Ibanez, J.F., Serrano Castaneda, J.E., Martinez Santos, J.C.: An IoT camera system for the collection of data using QR code as object recognition algorithm. In: 2018 Congreso Internacional de Innovacion y Tendencias en Ingenieria, CONIITI 2018—Proceedings, art. no. 8587087 (2018)
11. Babu, S., Markose, S.: IoT enabled Robots with QR code based localisation. In: 2018 International Conference on Emerging Trends and Innovations In Engineering And Technological Research, ICETIETR 2018, art. no. 8529028 (2018)
12. Narendar Singh, D., Anil Kumar, G., Sowjanya, K.: IOT based secured railway passengers service systems with QR code and unmanned railway crossing alarm. Int. J. Recent Technol. Eng. **7**(4), 283–289 (2018)
13. Linjie, L., Haijun, R.: The applied research on power telecommunication identifier management system based on QR code. In: Proceedings of the IEEE International Conference on Software Engineering and Service Sciences, ICSESS, pp. 270–274 (2018)
14. D'Addio, G., Smarra, A., Biancardi, A., Cesarelli, M., Arpaia, P.: Quick-response coding system for tracking rehabilitation treatments in clinical setting. In: 2017 IEEE International Workshop on Measurement and Networking, M and N 2017—Proceedings, art. no. 8078362 (2017)
15. Marktscheffel, T., Gottschlich, W., Popp, W., Werli, P., Fink, S.D., Bilzhause, A., De Meer, H.: QR code based mutual authentication protocol for internet of things. In: WoWMoM 2016–17th International Symposium on a World of Wireless, Mobile and Multimedia Networks, art. no. 7523562 (2016)
16. Zhu, N.X.: Application of QR code rebuild library integration community. Adv. Mater. Res. **805–806**, 1907–1910 (2013)
17. Gu, Y., Jiao, Y., Xu, X., Yu, Q.: Request-response and censoring-based energy-efficient decentralized change-point detection with IoT applications. IEEE Internet Things J
18. Zaman, M.U., Shen, T., Min, M.: Hash Vine: a New Hash structure for scalable generation of hierarchical hash codes. In: 2019 IEEE International Systems Conference (SysCon), Orlando, FL, USA, pp. 1–6 (2019)
19. Bittl, S.: Efficient construction of infinite length hash chains with perfect forward secrecy using two independent hash functions. In: 2014 11th International Conference on Security and Cryptography (SECRYPT), Vienna, pp. 1–8 (2014)
20. Fu, J., Wu, C., Wu, J., Fan, R., Ping, L.: Efficient hash chain traversal based on binary tree inorder traversal. In: 2010 Second Pacific-Asia Conference on Circuits, Communications and System, pp. 178–181, Beijing (2010)

21. Lin, P.Y., Chen, Y.H.: High payload secret hiding technology for QR codes. EURASIP J. Image Video Process, pp. 1–8 (2017)
22. Arebey, M., Hannan, M.A., Basri, H., Begum, R.A., Abdullah, H.: Solid waste monitoring system integration based on RFID, GPS and camera. In: 2010 International Conference on Intelligent and Advanced Systems, pp. 1–5, Manila, (2010)
23. Krithika, S., KajaMaideen, J., Madan, T.K.C.: An automated trash monitoring system for waste management using IoT-(I-Bin). Int. J. Eng. Adv. Technol. **8**(6 Special Issue 3), pp. 1134–1137 (2019)
24. Li, S., Zhou, B., Dai, J., Sun, X.: A secure scheme of continuity based on two-dimensional backward hash key chains for sensor networks. IEEE Wirel. Commun. Lett. **1**(5), 416–419 (2012)
25. Challal, Y., Bouabdallah, A., Hinard, Y.: Efficient multicast source authentication using layered hash-chaining scheme. In: 29th Annual IEEE International Conference on Local Computer Networks, pp. 411–412, Tampa, FL, USA (2004)
26. Wei, Z.: Self-updating hash chains based on erasure coding. In: 2010 International Conference on Computer, Mechatronics, Control and Electronic Engineering, pp. 173–175, Changchun (2010)
27. Yang, X., Wang, J., Chen, J., Pan, X.: A self-renewal hash chain scheme based on fair exchange idea (SRHC-FEI). In: 2010 3rd International Conference on Computer Science and Information Technology, pp. 152–156, Chengdu (2010)
28. Zhang, M.-Q., Dong, B., Yang, X.-Y.: A new self-updating hash chain structure scheme. In: 2009 International Conference on Computational Intelligence and Security, pp. 315–318, Beijing (2009)
29. Meng, J., Wang, Z.: A RFID security protocol based on hash chain and three-way handshake. In: 2013 International Conference on Computational and Information Sciences, pp. 1463–1466, Shiyang (2013)
30. Tewari, A., Gupta, B.B.: Security, privacy and trust of different layers in internet-of-things (IoTs) framework. Futur. Gener. Comput. Syst. **108**, 909–920 (2020)
31. Sahoo, S.R., Gupta, B.B.: Classification of spammer and nonspammer content in online social network using genetic algorithm-based feature selection. Enterp. Inf. Syst. **14**(5), 710–736 (2020)
32. Sahoo, S.R., Gupta, B.B.: Hybrid approach for detection of malicious profiles in twitter. Comput. Electr. Eng. **76**, 65–81 (2019)
33. Tergiou, C.L., Psannis, K.E., Gupta, B.B.: IoT-based big data secure management in the fog over a 6G wireless network. IEEE Internet of Things J. (2020)
34. Gupta, B.B., Quamara, M.: An overview of internet of things (IoT): architectural aspects, challenges, and protocols. Concurr. Comput.: Pract. Exp. **32**(21), e4946 (2020)
35. Benrhouma, O., Hermassi, H., Abd El-Latif, A.A., Belghith, S.: Chaotic watermark for blind forgery detection in images. Multimed. Tools Appl. **75**(14), 8695–8718 (2016)
36. Abd El-Latif, A.A., Abd-El-Atty, B., Mehmood, I., Muhammad, K., Venegas-Andraca, S.E., Peng, J., Quantum-inspired blockchain-based cybersecurity: securing smart edge utilities in IoT-based smart cities. Inf. Process. Manag. **58**(4), 102549 (2021)
37. Zhang, W.-Z., Elgendy, I.A., Hammad, M., Iliyasu, A.M., Du, X., Guizani, M., Abd El-Latif, A.A.: Secure and optimized load balancing for multi-tier IoT and edge-cloud computing systems. IEEE Internet Things J. (2020)
38. Abd El-Latif, A.A., Abd-El-Atty, B., Venegas-Andraca, S.E., Elwahsh, H., Piran, M.J., Bashir, A.K., Song, O.-Y., Mazurczyk, W.: Providing end-to-end security using quantum walks in IoT networks. IEEE Access **8**, 92687–92696 (2020)
39. Abd El-Latif, A.A., Abd-El-Atty, B., Hossain, M.S., Elmougy, S., Ghoneim, A.: Secure quantum steganography protocol for fog cloud internet of things. IEEE Access **6**, 10332–10340 (2018)
40. Jayan, A.P., Balasubramani, A., Kaikottil, A., Harini, N.: An enhanced scheme for authentication using OTP and QR code for MQTT protocol. Int. J. Recent Technol. Eng. **7**(5), 70–75 (2019)

Data Security Challenges in Deep Neural Network for Healthcare IoT Systems

Edmond S. L. Ho

Abstract With the advancement of IoT technology, more and more healthcare applications were developed in recent years. In addition to the traditional sensor-based systems, image-based healthcare IoT systems become more popular since no specialized sensors are required. Combining with Deep Neural Network (DNN) based automated diagnosis and decision-making systems, it is possible to provide users with 24/7 health monitoring in real life. However, the high computational cost for training DNNs can be a hurdle for developing such kind of powerful systems. While cloud computing can be a feasible solution, uploading training data for the DNN models to the cloud may lead to data security issues. In this chapter, we will review some image-based healthcare IoT systems and discuss some potential risks on data security when training the DNN models on the cloud.

1 Introduction

Nowadays, portal devices with high computational power, as well as high-speed data transmission networks [18], are more accessible and affordable to the general public. This certainly enables a wide range of serious applications to be developed to improve our quality of life. In particular, such as environment is suitable for creating healthcare IoT applications which usually consist of capturing health-related data using sensors from the end-user. The data will then be uploaded to the cloud for (1) automated analysis and diagnosis, and/or (2) informing the medical experts or carers in a timely manner if abnormal health conditions are identified.

Examples of healthcare IoT systems in the literature include [25] in which heartbeat sensor, body temperature sensor, room temperature sensor, CO sensor, and CO_2 sensor were used for capturing health-related data in a hospital environment to serve as basic health signs monitoring system for patients. Chatterjee et al. [7] proposed

E. S. L. Ho (✉)
Northumbria University, Sutherland Building, Northumbria University,
Sutherland Building, NE1 8ST Newcastle upon Tyne, UK
e-mail: e.ho@northumbria.ac.uk

© The Author(s), under exclusive license to Springer Nature Switzerland AG 2022
A. A. Abd El-Latif et al. (eds.), *Security and Privacy Preserving for IoT and 5G Networks*,
Studies in Big Data 95, https://doi.org/10.1007/978-3-030-85428-7_2

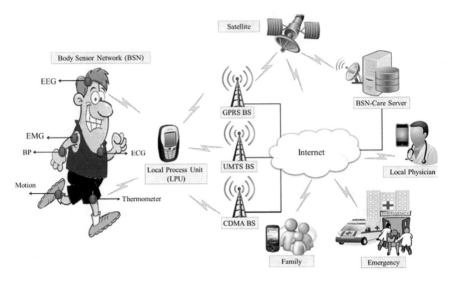

Fig. 1 The secure healthcare IoT system using body sensor network (BSN) technology illustrated in [17]. Image reproduced from [17]

an IoT system for cardiovascular diseases risk assessment of the user based on physiological parameters (age, gender, systolic and diastolic blood pressure, cholesterol, diabetes and smoking habits). Gope and Hwang [17] highlighted the need for a secure healthcare IoT system and proposed BSN-Care which is a secure IoT system based on Body Sensor Network (BSN) technology [30]. In particular, a wide range of sensors was used such a system, including Electrocardiogram (ECG), Electromyography (EMG), Electroencephalography (EEG), Blood Pressure (BP), etc. The secure IoT system is illustrated in Fig. 1. In addition, both network security (e.g. secure localization, authentication and anonymity) and data security (e.g. data integrity, data freshness and data privacy) are considered when designing the system [1–5, 13, 20, 21, 39, 40].

Data protection is an important aspect in IoT systems [45] which is not only specific to healthcare applications, but also all other systems that contain and transfer sensitive personal information. For healthcare applications, however, data protection and security become more important due to the fact that highly sensitive personal information will be processed and transferred within the IoT systems. In particular, encryption-based techniques have been widely used for protecting sensitive data to be sent over the network. In [12], Elhoseny et al. proposed a hybrid encryption system for securing secret medical data (such as the diagnostic results). The proposed framework is illustrated in Fig. 2. Specifically, the encryption method is based on a wide range of methods including Advanced Encryption Standard, Rivest, Shamir, and Adleman algorithms, and the method hides the secret data in a cover image.

In this chapter, we will focus on a less researched area. Firstly, we will review the related research in image-based healthcare IoT systems (Sect. 2) which consists

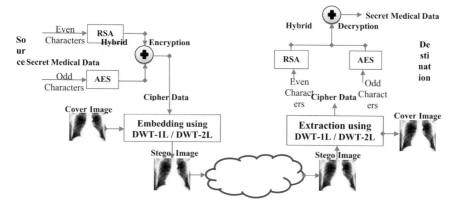

Fig. 2 The framework for securing medical data transmission in [12]. Image reproduced from [12]

of Deep Neural Network (DNN) as the core technology for automated diagnosis or decision making. Next, different methods for attacking DNNs will be discussed in Sect. 3.

2 Image-Based IoT Systems for Automated Diagnosis

Analyzing retinal images provide significant values to the healthcare sector since a wide range of health disorders, such as atherosclerosis, diabetic retinopathy, and congestive heart failure, can be predicted from those images. Recent work by Poplin et al. [36] demonstrated the results on predicting the cardiovascular risk factors from a large retinal fundus photographs dataset, which includes retinal fundus images from 48,101 patients from the UK Biobank (http://www.ukbiobank.ac.uk/about-biobank-uk) and 236,234 patients from EyePACS (http://www.eyepacs.org). The proposed models were validated using images from 12,026 patients from the UK Biobank and 999 patients from EyePACS. In [10], Das R. et al. proposed a low cost and portal healthcare conceptual framework for acquiring retinal fundus images using a Head Mounted Device (HMD). By combining such a framework with a mobile App for Internet connection and deep learning for automated diagnosis, this can potentially be a practical solution for developing countries that have limited access to low-cost retinal image acquisition. In particular, the HMD conceptual image is illustrated in Fig. 3. The linear actuator will automatically adjust the length to ensure the correct distance between the eyes and lenses can be maintained when capturing the retinal fundus images. Next, the acquired images will be sent to the automated diagnosis system which is essentially a DNN-based image classification framework (such as [36]) as cloud services. Finally, the predicted results will be sent to the medical doctors and/or experts for further analysis. At the time of publishing [10], the system is being translated into a smart healthcare application.

Fig. 3 The conceptual illustration of the proposed head mounted device (HMD) for capturing retinal fundus images in [10]. Image reproduced from [10]

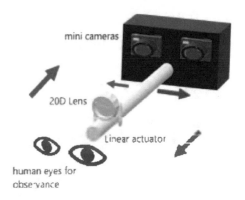

Nowadays, the generality of IoT enables a wide range of smart applications to be developed by connecting different types of sensors to the Internet which provide data to be analyzed by automated systems and notify human experts in a timely manner. However, data protection and privacy of the users are crucial factors to ensure the applications are being safe to use. Data encryption is a natural solution for this problem by protecting the data being read by unauthorized parties, but the additional computation cost can greatly degrade the performance of the IoT system especially for systems required to process high volume data. In [27], Jiang et al. proposed an efficient framework for encrypting and diagnosing Diabetic Retinopathy (DR) from retinal images using a camera sensor connected to a Raspberry Pi as a healthcare IoT application. To reduce the computational cost for data encryption, somewhat homomorphic encryption (SHE) is used. Furthermore, parallel homomorphic evaluation is performed by packing multibits into a single ciphertext through single instruction multiple data (SIMD). By this, the computation time for homomorphic evaluation and the transmission time for the cipher text can be reduced simultaneously. For diagnosing DR, density-based clustering [47] is used to classify the retinal images in a highly efficient manner. In summary, the proposed architecture provides a practical solution as a healthcare system with considerations on computational efficiency, privacy protection, storage overhead and communication cost (Fig. 4).

Liu et al. [32] proposed combining IoT and artificial intelligence for dental healthcare. In particular, an image-based dental health analysis system is implemented and evaluated. The overview of the proposed framework is illustrated in Fig. 5. Such an architecture provides users with an in-home dental healthcare platform for sending color dental images (i.e. RGB images) using a mobile terminal (e.g. smartphone) to the *Smart Dental Services Layer* for detecting dental diseases using a deep learning framework. The *AI Diagnosis of the Teeth* framework is trained using image data annotated with 7 types of dental diseases, including dental caries, dental uorosis, periodontal disease, cracked tooth, dental calculus, dental plaque, and tooth loss. A total of 12,600 clinical images were collected as training samples from 10 private clinics. The core of the AI diagnosis framework is based on Mask R-CNN [22], and image enhancement (such as balancing the dynamic range, edge, and color) is

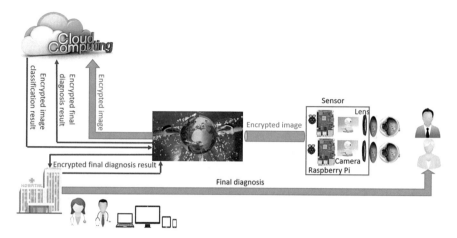

Fig. 4 The overview of the proposed framework in [27]. Image reproduced from [27]

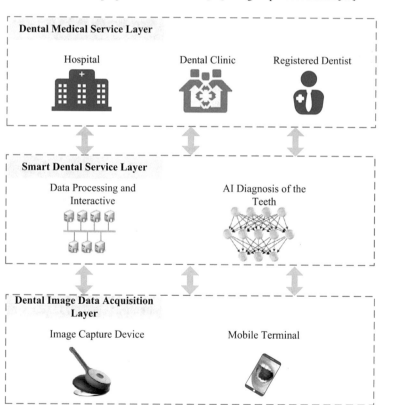

Fig. 5 The overview of the proposed framework in [32]. Image reproduced from [32]

required to achieve better performance. Experimental results show that the diagnosis accuracy ranged from 90.1 to 100% for the 7 types of dental diseases and the mean diagnosis time is reduced by 37.5% in the 1-month trial period in 10 private clinics.

In addition to the aforementioned specialized medical photography IoT systems, there is also a huge potential for typical RGB image (e.g. captured using smartphones, camcorders, etc) based IoT systems to be developed as healthcare applications. For example, RGB videos (or RGB image sequences) can be used for predicting cerebral palsy through clinical assessments such as General Movements Assessment (GMA) [11]. The main focus of GMA is assessing the complexity and variability of the general movements of the infant to predict if the nervous system is impaired or not. Most of the current clinical practise still rely on highly trained clinicians to inspect the RGB videos manually. This opens the door for automating this process to:

- reduce subjectivity on the manual assessment
- improve efficiency
- monitor (24/7) the high-risk group such as infants born preterm [19].

A recent work proposed by McCay et al. [33] automate GMA by using computer vision and machine learning techniques. Specifically, skeletal poses are extracted from the video recordings of infant body movements in the pre-processing stage. An example of the skeletal pose estimated from a video frame using OpenPose [6] is illustrated in Fig. 6. Based on the skeletal poses which demonstrated encouraging results in gait analysis [38] as a healthcare application, two new pose-based features, namely Histograms of Joint Orientation 2D (HOJO2D) and Histograms of Joint Displacement 2D (HOJD2D), are proposed to facilitate the classification process to predict if the movement of the infant is considered as *normal* or *abnormal*. In particular, HOJO2D represents the distribution of the orientations of the body parts while HOJD2D represents the distribution of the joint velocity. By fusing such features extracted at each joint, the fused features can be used for representing movement at different levels such as joint-level, limb-level and full body-level. With the use of traditional classifiers such as k-nearest neighbor, Support Vector Machine (SVM) and Ensemble classifier, encouraging result with 91.67% classification accuracy was obtained.

To further improve the classification accuracy, McCay et al. extended the work [34] by proposing 5 DNN architectures as shown in Fig. 7. Specifically, the new network architectures include fully-connected networks (Fig. 7a), 1D convolutional neural networks (Fig. 7b, c) and 2D convolutional neural networks (Fig. 7d, e). Experimental results highlighted that the proposed DNN classifiers are more robust than the traditional classifiers evaluated in [33], and the 1D convolutional neural networks achieved the best performance with the HOJO2D and HOJD2D features (Fig. 8).

While McCay et al. [33, 34] mainly focus on the automated diagnostic framework, such an approach can be extended as an IoT application for healthcare by providing the automated diagnosis as cloud services such as the architecture illustrated in Fig. 9. The use of skeletal pose sequence for the cerebral palsy prediction can certainly lower the risk of leaking sensitive data in case of a data breach since it will be

Fig. 6 An example of skeletal pose estimation results presented in [33]. Image reproduced from [33]

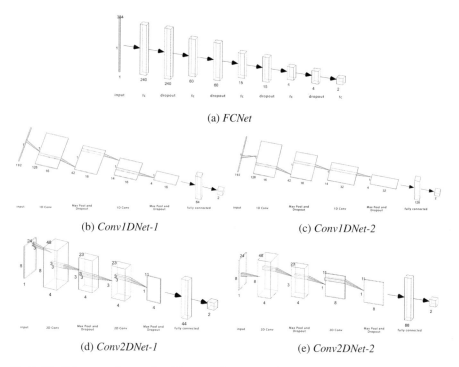

Fig. 7 The 5 deep neural network architectures for automated prediction of cerebral palsy proposed in [34], including the fully-connected layers based *FCNet*, 1D convolutional neural network based *Conv1DNet-1* and *Conv1DNet-2*, and 2D convolutional neural Network based *Conv2DNet-1* and *Conv2DNet-2*. Images reproduced from [34]

Fig. 8 An example of skeletal pose estimation results obtained using TensorFlow lite PoseNet. Image reproduced from https://www.tensorflow.org/lite/examples

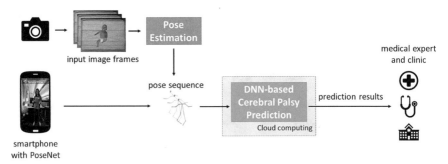

Fig. 9 Extending the pose-based cerebral palsy prediction framework to a healthcare IoT system

difficult to trace the identity of the subject by only looking at the skeletal poses. To further enhance the accessibility, the video capturing and pose estimation can be implemented as a smartphone App such as the TensorFlow Lite PoseNet (https://www.tensorflow.org/lite/examples, see Fig. 8). This can possibly be used as in-home monitoring IoT system.

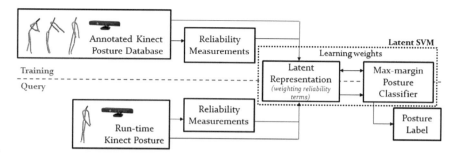

Fig. 10 The posture monitoring framework proposed in [23]. Image reproduced from [23]

Pose analysis can also be used in a smart office environment. In [23], Ho et al. proposed an RGB-D camera based monitoring framework (Fig. 10) to assess the healthiness of the postures of the user in an office environment. Examples of healthy and unhealthy postures are illustrated in Fig. 11. With the advancement of depth-sensing technology, RGB-D cameras such as Microsoft Kinect become more and more affordable to make it feasible to incorporate such devices in IoT systems. However, the 2.5D data captured from Microsoft Kinect can be noisy and result in incorrect pose extraction (see Fig. 12). To tackle this problem, Ho et al. [23] proposed to compute an extended set of *reliability values* [41] of the detected joint locations from Microsoft Kinect and incorporate those values into the classification framework. By this, the reliability values will determine the importance of each detected joint locations when classifying the input postures into different healthy/unhealthy classes. Experimental results show that the proposed method outperformed the baselines in the study.

The aforementioned IoT systems take image or image sequence as input for auto-mated diagnosis. Therefore, the quality of the images is crucial to the success of the systems. However, image quality can be difficult to control when the data is captured at the user-end in which the ideal environment (such as lighting condition, white balance, etc) is not available and the users are inexperienced in taking pictures. To tackle this problem, More et al. [35] proposed a secure Internet of Healthcare Things (IoHT) system which consists of a sparse aware with convolution neural network (SA_CNN) for effective noise removal (Fig. 13a) and a secure IoHT architecture for medical data storage (Fig. 13b). The proposed framework was evaluated using var-ious medical modalities and the experimental results show that the new framework outperformed the related work [8, 9, 14, 50] on quantitative measurements such as peak signal to noise ratio (PSNR), structural similarity index (SSIM), and mean squared error (MSE).

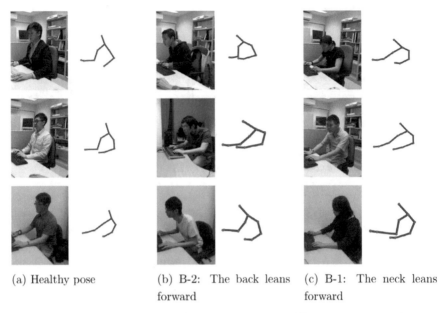

(a) Healthy pose

(b) B-2: The back leans forward

(c) B-1: The neck leans forward

Fig. 11 Examples of health and unhealthy postures collected in [23]. Image reproduced from [23]

Fig. 12 Examples of joint locations detected using the microsoft Kinect SDK v1. It can be seen that the joint locations (such as the elbow location in the middle column) are detected incorrectly. Image reproduced from [23]

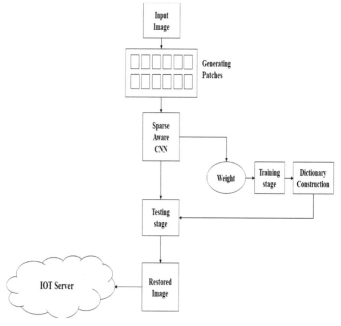

(a) The SA_CNN architecture

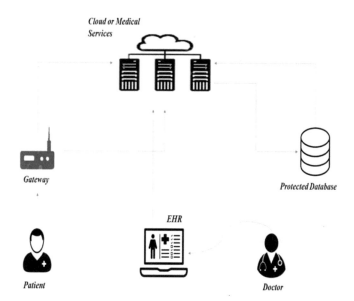

(b) The IoHT architecture for medical data storage

Fig. 13 The sparse aware convolution Neural Network (SA_CNN) architecture and IoHT architecture for medical data storage proposed in [35]. Images reproduced from [35]

3 Data Security Issues in Image-Based Deep Learning

With the outstanding performance in using deep learning in different research areas, more and more real-world applications start taking advantage of incorporating DNNs such as the IoT systems discussed in Sect. 2. However, the training process of DNNs usually requires a high volume of training data as well as high computational costs. As a result, cloud service providers including Google, Amazon and Microsoft are offering cloud-based deep learning solutions which can be referred to as Machine Learning as a Service (MLaaS). While such kind of services provides users with the flexibility on the computational resources (i.e. computational power and storage space), uploading training data to remote servers which are managing by external companies may lead to data security problems. Xu et al. [48] recently reviewed some of the data security issues and the solutions available currently, as well as proposed a verifiable and privacy-preserving prediction protocol, namely *SecureNet* for protecting the deep neural networks model and user privacy. In particular [48], focusing on typical DNN training processing depicted in Fig. 14 which has an input layer, a number of hidden layers and an output layer. In particular, the output layer contains the prediction (e.g. class label for classification problems). The data security issues are mainly related to attacking the DNN model training process such that wrong prediction will be given as output. In the rest of this section, the terminology from [48] will be used for explaining different types of attacks and solutions.

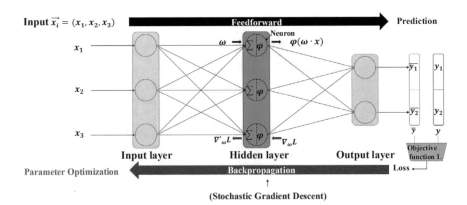

Fig. 14 A general deep neural network training process illustrated in [48]. Image reproduced from [48]

3.1   Terminology

3.1.1   Centralized Training

This refers to the training process will be done on a single server. Here a single cloud server will be used for getting all training data from the user to complete the DNN training process. In other words, the parameters of the resultant DNN model are computed from the cloud server solely.

3.1.2   Collaborative Training

This refers to the training process will be done in a distributed manner. Each user will train the DNN model separately while exchanging the learned model parameters. By this, the 'final' DNN model is trained in a 'collaborative' manner.

3.1.3   Black-Box Attacker

The attacker can access the DNN model to generate the output (i.e. prediction). However, the attacker does not know the details of the DNN model such as the training data, network architecture and optimization procedures.

3.1.4   White-Box Attacker

The attacker can access the DNN model to generate the output (i.e. prediction) as well as the details of the DNN model such as the training data, network architecture and optimization procedures.

3.2   Poisoning Attack

The first type of attacks reviewed in [48] is the Poisoning attack. The idea is to have 'poisoned data' in the training dataset such that that DNN model to be learned will likely generate the wrong prediction. As illustrated in Fig. 14, the parameters of the DNN model is learned during *backpropagation* in which the gradient is updated in order to minimize the *loss* term(s) of the model. It can be seen that if the training data is being poisoned (or manipulated), the DNN model parameters will not be learned correctly and it may lead to the wrong prediction.

In [26], Jagielski et al. presented a study on poisoning attack against linear regression (Fig. 15) and proposed an effective defence method, namely *TRIM*, for those attacks. The poisoning attack is done by introducing *poisoning points* into the training

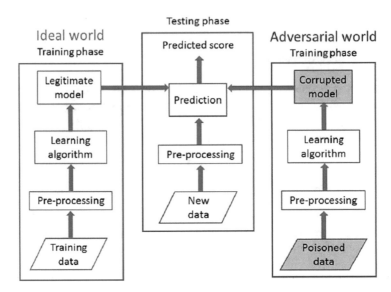

Fig. 15 The system architecture for simulating the 'normal' (Ideal world) and 'attack' (Adversarial world) scenarios in [26]. Image reproduced from [26]

data. In particular, both white-box and black-box attackers scenarios were considered and the problems were formulated as a generic bilevel optimization problem. Specifically, for white-box attackers, the outer optimization is responsible for selecting the poisoning points to maximize the loss (opposite to the typical training process which minimizes the loss) while the inner optimization is used for training the regression model parameters based on the poisoned data manipulated by the outer optimization. The black-box attacks follow the same formulation except the training data has to be prepared by the attacker instead of using the 'real' training data.

To defend against the poisoning attacks, the proposed TRIM algorithm [26] not only ignore (or remove) outliers (i.e. likely to be the poisoning points) from the training data as in previous work, but also ignore *inliers* which are the poisoning points that are having a similar data distributions as the real training data. This is achieved by iterative updating the regression model parameters using a subset of training data that has the lowest residuals. Experimental results show that the TRIM algorithm can effectively defence poisoning attacks as the mean square errors (MSEs) are within 1% when compared with the MSEs from the unpoisoned models.

Suciu et al. [42] proposed the *FAIL* framework for evaluating the robustness of machine learning models against poisoning and evasion attacks along 4 dimensions: Features, Algorithms, Instances, and Leverage. The framework evaluates a machine learning model by treating different knowledge levels (i.e. availability of model details as in black-box and white-box attacks) as different scenarios and returns the success rates of different attacks. An attack algorithm called *StingRay* is further proposed which demonstrated the effectiveness of bypassing two existing anti-poisoning defences. The main idea behind StingRay is to introduce poisoning points while not affecting the overall classification performance of the DNN model. Specifically, given a target class of features to be misclassified as the goal of the poisoning attack, StingRay will create poisoning samples that are based on a benign sample from the training data. Next, a subset of the base sample will be replaced by the features of the target class. This process will repeat until the target class of features is being misclassified.

3.3 Evasion Attack

The second type of attacks reviewed in [48] is the Evasion attack, in which the goal of the attack is to input carefully crafted samples (i.e. *adversarial examples*) during the prediction (or testing) stage and lead to misclassification. In particular, the adversarial examples can be classified correctly by human but not DNN models. In Fig. 16, Goodfellow et al. [15] demonstrated an adversarial example which led to misclassification. Specifically, adversarial examples can be generated by adding noise (in [15], the perturbation is based on the gradient of the cost function) to the original input data (e.g. the 'original' panda image in Fig. 16 left). The resultant adversarial example (Fig. 16 right) will appear similar to the original image to human eyes, but such input will lead to wrong classification result in DNN models such as classifying this 'panda' image as the 'gibbon' class by GoogLeNet [43].

In the rest of this section, some popular techniques for generating adversarial samples will be discussed. Readers are referred to a recent systematic review [49] as well as the original publications for the details.

Szegedy et al. [44] investigated different properties of neural networks. From their experiments, they found that input-output mappings learned by DNNs are discontinuous. Based on such an interesting property, they proposed an optimization

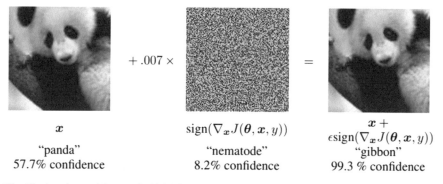

$$x$$
"panda"
57.7% confidence

$$\text{sign}(\nabla_x J(\theta, x, y))$$
"nematode"
8.2% confidence

$$x + \epsilon \text{sign}(\nabla_x J(\theta, x, y))$$
"gibbon"
99.3 % confidence

Fig. 16 An adversarial example (right) is created by perturbating the original image (left) based on the gradient of the cost function in [15]. The adversarial example leads to misclassification result on GoogLeNet [43]. Image reproduced from [15]

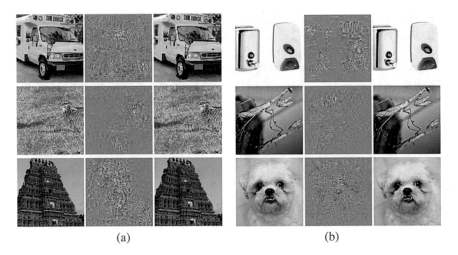

(a)   (b)

Fig. 17 An adversarial examples generated for AlexNet [28] using the method proposed in [44]. In each set of samples (**a** and **b**) the images on the left column are the original images while the right column contains the adversarial examples which lead to misclassification. The middle column illustrates the differences between the original image and the adversarial example magnified by 10 times. Image reproduced from [44]

technique to generate adversarial samples. Specifically, the optimization search for the required perturbation by minimizing the probability of images being classified as the correct class. Adversarial samples were generated successfully for attacking various networks (MNIST, QuocNet [31], AlexNet [28]) in the study. Adversarial examples generated for AlexNet [28] are shown in Fig. 17. It can be seen that the original images and adversarial examples are indistinguishable by a human, but DNN models will misclassify the adversarial examples.

Goodfellow et al. [15] proposed a method called fast gradient sign method (FGSM) which is more computationally efficient than the expensive optimization used in [44]. Specifically, an adversarial example x' can be generated adding perturbation η to the input image x based on the sign of the gradient of the cost function in the training process:

$$\eta = \epsilon \, \text{sign}(\nabla_x J(\theta, x, y)) \tag{1}$$

where y is the target associated with x, θ is the model parameters, $J(\theta, x, y)$ is the cost of training the network and ϵ is used for controlling the strength of the perturbation. An example of the adversarial example generated by this method is visualized in Fig. 16. Since then, there are variants of the FGSM method proposed in the literature. For example, Tramèr et al. [46] introduced randomness when perturbing the input data, Rozsa et al. [37] replaced the sign of the gradient with the raw gradient value, and Kurakin et al. [29] further extend FGSM to maximizing the target class probability while generating adversarial examples.

4 Conclusion

In this chapter, we reviewed a wide range of IoT-based healthcare systems. With the advancement of technology, such as more affordable and portal cameras and better performance smartphones, it is now feasible to capture images from the user-end using portable devices in healthcare IoTs. Combining with image-based deep learning frameworks on the cloud for diagnosis, fully or semi-automated IoT solutions can be realized. While this opens the door for a wide variety of image-based healthcare IoTs to be utilized in real life, we also review some of the potential risks for the image-based deep learning framework which may lead to making the wrong decision. It is advised that IoT solution developers should understand more about the risks and counter those attacks accordingly. For example, Goodfellow et al. [16] and Huang et al. [24] improve the robustness of the DNN models by adding a small number of adversarial examples to the training set iterative. By this, the model to be trained will be more robust again the adversarial examples that may be injected by attackers during the prediction stage.

Acknowledgements This project is supported by the Royal Society (Ref: IES/R1/191147).

References

1. Abd-El-Atty, B., Iliyasu, A.M., Alaskar, H., El-Latif, A., Ahmed, A.: A robust quasi-quantum walks-based steganography protocol for secure transmission of images on cloud-based e-healthcare platforms. Sensors **20**(11), 3108 (2020)
2. Abd EL-Latif, A.A., Abd-El-Atty, B., Abou-Nassar, E.M., Venegas-Andraca, S.E.: Controlled alternate quantum walks based privacy preserving healthcare images in internet of things. Opt. Laser Technol. **124**, 105942 (2020)
3. Abd El-Latif, A.A., Hossain, M.S., Wang, N.: Score level multibiometrics fusion approach for healthcare. Cluster Comput. **22**(1), 2425–2436 (2019)
4. Abou-Nassar, E.M., Iliyasu, A.M., El-Kafrawy, P.M., Song, O.-Y., Bashir, A.K., Abd El-Latif, A.A.: Ditrust chain: towards blockchain-based trust models for sustainable healthcare IoT systems. IEEE Access **8**, 111223–111238 (2020)
5. Alghamdi, A., Hammad, M., Ugail, H., Abdel-Raheem, A., Muhammad, K., Khalifa, H.S., Abd El-Latif, A.A.: Detection of myocardial infarction based on novel deep transfer learning methods for urban healthcare in smart cities. Multimed. Tools Appl. 1–22 (2020)
6. Cao, Z., Martinez, G.H., Simon, T., Wei, S., Sheikh, Y.A.: Openpose: realtime multi-person 2D pose estimation using part affinity fields. IEEE Trans. Pattern Anal. Mach. Intell. (2019)
7. Chatterjee, P., Cymberknop, L.J., Armentano, R.L.: IoT-based decision support system for intelligent healthcare applied to cardiovascular diseases. In: 2017 7th International Conference on Communication Systems and Network Technologies (CSNT), pp. 362–366 (2017)
8. Chen, Y., Pock, T.: Trainable nonlinear reaction diffusion: a flexible framework for fast and effective image restoration. IEEE Trans. Pattern Anal. Mach. Intell. **39**(6), 1256–1272 (2017)
9. Dabov, K., Foi, A., Katkovnik, V., Egiazarian, K.: Image denoising by sparse 3-D transform-domain collaborative filtering. IEEE Trans. Image Process. **16**(8), 2080–2095 (2007)
10. Das R., Rakshitha, G., Juvanna, I., Subramanian, D.V.: Retinal based automated healthcare framework via deep learning. In: 2018 Second International Conference on Green Computing and Internet of Things (ICGCIoT), pp. 93–97 (2018)

11. Einspieler, C., Prechtl, H.F.R.: Prechtl's assessment of general movements: a diagnostic tool for the functional assessment of the young nervous system. Mental Retardation Dev. Disabil. Res. Rev. **11**(1), 61–67 (2005)
12. Elhoseny, M., Ramilez, G., Abu-Elnasr, O.M., Shawkat, S.A., Arunkumar, N., Farouk, A.: Secure medical data transmission model for IoT-based healthcare systems. IEEE Access **6**, 20596–20608 (2018)
13. Gad, R., Talha, M., Abd El-Latif, A.A., Zorkany, M., Ayman, E.-S., Nawal, E.-F., Muhammad, G.: Iris recognition using multi-algorithmic approaches for cognitive internet of things (CIOT) framework. Future Gener. Comput. Syst. **89**, 178–191 (2018)
14. Gai, S., Bao, Z.: New image denoising algorithm via improved deep convolutional neural network with perceptive loss. Expert Syst. Appl. **138**, 112815 (2019)
15. Goodfellow, I., Shlens, J., Szegedy, C.: Explaining and harnessing adversarial examples. In: International Conference on Learning Representations (2015)
16. Goodfellow, I.J., Shlens, J., Szegedy, C.: Explaining and harnessing adversarial examples (2015)
17. Gope, P., Hwang, T.: BSN-care: a secure IoT-based modern healthcare system using body sensor network. IEEE Sens. J. **16**(5), 1368–1376 (2016)
18. Gupta, B., Quamara, M.: An overview of internet of things (IoT): architectural aspects, challenges, and protocols. In: Concurr. Comput.: Pract. Exp. **32**(21), e4946 CPE-18-0159.R1 (2020)
19. Hafström, M., Källén, K., Serenius, F., Maršál, K., Rehn, E., Drake, H., Ådén, U., Farooqi, A., Thorngren-Jerneck, K., Strömberg, B.: Cerebral palsy in extremely preterm infants. Pediatrics **141**(1) (2018)
20. Hammad, M., Alkinani, M.H., Gupta, B., Abd El-Latif, A.A.: Myocardial infarction detection based on deep neural network on imbalanced data. Multimed. Syst. 1–13 (2021)
21. Hammad, M., Iliyasu, A.M., Subasi, A., Ho, E.S., Abd El-Latif, A.A.: A multitier deep learning model for arrhythmia detection. IEEE Trans. Instrum. Meas. **70**, 1–9 (2020)
22. He, K., Gkioxari, G., Dollár, P., Girshick, R.B.: Mask R-CNN. CoRR (2017). arXiv:abs/1703.06870
23. Ho E.S., Chan, J.C., Chan, D.C., Shum, H.P., Cheung, Y.M., Yuen, P.C.: Improving posture classification accuracy for depth sensor-based human activity monitoring in smart environments. Comput. Vis. Image Underst. 97–110 (2016)
24. Huang, R., Xu, B., Schuurmans, D., Szepesvari., C.: Learning with a strong adversary (2016)
25. Islam, M., Rahaman, A., Islam. R.: Development of smart healthcare monitoring system in iot environment. SN Comput. Sci. **1**, 185 (2020)
26. Jagielski, M., Oprea, A., Biggio, B., Liu, C., Nita-Rotaru, C., Li, B.: Manipulating machine learning: Poisoning attacks and countermeasures for regression learning. In: 2018 IEEE Symposium on Security and Privacy (SP), pp. 19–35 (2018)
27. Jiang, L., Chen, L., Giannetsos, T., Luo, B., Liang, K., Han, J.: Toward practical privacy-preserving processing over encrypted data in IoT: an assistive healthcare use case. IEEE Internet Things J. **6**(6), 10177–10190 (2019)
28. Krizhevsky, A., Sutskever, I., Hinton, G.E.: Imagenet classification with deep convolutional neural networks. In: Pereira, F., Burges, C.J.C., Bottou, L., Weinberger, K.Q. (eds.) Advances in Neural Information Processing Systems, vol. 25, pp. 1097–1105. Curran Associates, Inc., (2012)
29. Kurakin, A., Goodfellow, I.J., Bengio, S.: Adversarial machine learning at scale. In: International Conference on Learning Representations (2017)
30. Lai, X., Liu, Q., Wei, X., Wang, W., Zhou, G., Han, G.: A survey of body sensor networks. Sensors **13**(5), 5406–5447 (2013)
31. Le, Q.V., Ranzato, M., Monga, R., Devin, M., Chen, K., Corrado, G.S., Dean, J., Ng, A.Y.: Building high-level features using large scale unsupervised learning. In: Proceedings of the 29th International Coference on International Conference on Machine Learning, ICML'12, pp. 507–514, Madison, WI, USA (2012). Omnipress
32. Liu, L., Xu, J., Huan, Y., Zou, Z., Yeh, S.C., Zheng, L.R.: A smart dental health-IoT platform based on intelligent hardware, deep learning, and mobile terminal. IEEE J. Biomed. Health Inform. **24**(3), 898–906 (2020)

33. McCay, K.D., Ho, E.S.L., Marcroft, C., Embleton, N.D.: Establishing pose based features using histograms for the detection of abnormal infant movements. In: 2019 41st Annual International Conference of the IEEE Engineering in Medicine and Biology Society (EMBC), pp. 5469–5472 (2019)
34. McCay, K.D., Ho, E.S.L., Shum, H.P.H., Fehringer, G., Marcroft, C., Embleton, N.D.: Abnormal infant movements classification with deep learning on pose-based features. IEEE Access **8**, 51582–51592 (2020)
35. More, S., Singla, J., Kavita, S.V., Ghosh, U., Rodrigues, J.J.P.C., Hosen, A.S.M.S., Ra, I.: Security assured CNN-based model for reconstruction of medical images on the internet of healthcare things. IEEE Access **8**, 126333–126346 (2020)
36. Poplin, R., Varadarajan, A.V., Blumer, K., Liu, Y., McConnell, M., Corrado, G., Peng, L., Webster, D.: Predicting cardiovascular risk factors in retinal fundus photographs using deep learning. Nat. Biomed. Eng. (2018)
37. Rozsa, A., Rudd, E.M., Boult, T.E.: Adversarial diversity and hard positive generation. In: 2016 IEEE Conference on Computer Vision and Pattern Recognition Workshops (CVPRW), pp. 410–417 (2016)
38. Rueangsirarak, W., Zhang, J., Aslam, N., Ho, E.S.L., Shum, H.P.H.: Automatic musculoskeletal and neurological disorder diagnosis with relative joint displacement from human gait. IEEE Trans. Neural Syst. Rehabil. Eng. **26**(12), 2387–2396 (2018)
39. Sedik, A., Hammad, M., Abd El-Samie, F.E., Gupta, B.B., Abd El-Latif, A.A.: Efficient deep learning approach for augmented detection of coronavirus disease. Neural Comput. Appl. 1–18 (2021)
40. Sedik, A., Iliyasu, A.M., El-Rahiem, A., Abdel Samea, M.E., Abdel-Raheem, A., Hammad, M., Peng, J., El-Samie, A., Fathi, E., El-Latif, A.A.A., et al.: Deploying machine and deep learning models for efficient data-augmented detection of COVID-19 infections. Viruses **12**(7), 769 (2020)
41. Shum, H.P.H., Ho, E.S.L., Jiang, Y., Takagi, S.: Real-time posture reconstruction for microsoft kinect. IEEE Trans. Cybern. **43**(5), 1357–1369 (2013)
42. Suciu, O., Marginean, Kaya, Y., III, H.D., Dumitras, T.: When does machine learning FAIL? generalized transferability for evasion and poisoning attacks. In: 27th USENIX Security Symposium (USENIX Security 18), pp. 1299–1316, Baltimore, MD. USENIX Association (2018)
43. Szegedy, C., Liu, W., Jia, Y., Sermanet, P., Reed, S., Anguelov, D., Erhan, D., Vanhoucke, V., Rabinovich, A.: Going deeper with convolutions. In: 2015 IEEE Conference on Computer Vision and Pattern Recognition (CVPR), pp. 1–9 (2015)
44. Szegedy, C., Zaremba, W., Sutskever, I., Bruna, J., Erhan, D., Goodfellow, I., Fergus, R.: Intriguing properties of neural networks (2014)
45. Tewari, A., Gupta, B.: Security, privacy and trust of different layers in internet-of-things (IoTs) framework. Future Gener. Comput. Syst. **108**, 909–920 (2020)
46. Tramèr, F., Kurakin, A., Papernot, N., Goodfellow, I., Boneh, D., McDaniel, P.: Ensemble adversarial training: attacks and defenses. In: International Conference on Learning Representations (2018)
47. Tron, R., Zhou, X., Esteves, C., Daniilidis, K.: Fast multi-image matching via density-based clustering. In: 2017 IEEE International Conference on Computer Vision (ICCV), pp. 4077–4086 (2017)
48. Xu, G., Li, H., Ren, H., Yang, K., Deng, R.H.: Data security issues in deep learning: attacks, countermeasures, and opportunities. IEEE Commun. Mag. **57**(11), 116–122 (2019)
49. Yuan, X., He, P., Zhu, Q., Li, X.: Adversarial examples: attacks and defenses for deep learning. IEEE Trans. Neural Netw. Learn. Syst. **30**(9), 2805–2824 (2019)
50. Zhang, K., Zuo, W., Chen, Y., Meng, D., Zhang, L.: Beyond a gaussian denoiser: residual learning of deep CNN for image denoising. IEEE Trans. Image Process. **26**(7), 3142–3155 (2017)

Efficacious Data Transfer Accomplished by Trustworthy Nodes in Cognitive Radio

Avila Jayapalan, Padmapriya Praveenkumar, Prem Savarinathan, and Thenmozhi Karuppasamy

Abstract Cognitive Radio (CR) utilizes the free spectrum through a process called spectrum sensing efficiently. This free spectrum is shared by many CR's in the wireless environment. Many CR can shape a network and utilize the resources. A cognitive node can dynamically be part of the cluster or organization. The members of the network are managed and monitored by master controller called Network Manager (NM). To reduce the burden of NM sub-controller called Group Head (GH) are assigned. Suppose if a CR wish to become member of the network, then it is the responsibility of the NM and GH to check the reliability of the CR and then decide whether to accept it or not as the member of the network. With the above back ground this work focusses on two things. First one is to check the trustworthiness of the CR and to decide whether the cognitive radio can end up as a member of the group so that it can get access to all the features available in that group. This is done by asking many questions to the CR by the NM and GH. To overcome the malicious attack in the wireless medium the communication between CR and NM and also the communication between CR and GH are encrypted. Here Rijndael AES encryption algorithm is used to encrypt the questions. The key to the algorithm is generated using sutras wheel. Once the CR becomes the member of the CR network it is assigned with a free channel if available. Utilizing this channel, the CR can transmit the data to its intended receiver. To protect the data from intruders the information is encrypted. Here DNA algorithm used to encrypt the information to be transmitted and the same is decrypted at its destination. The dynamic spectrum access and security information transfer via the free channel integrated together makes the wireless communication the most successful one.

Keywords Cognitive radio · NM · GH · Bi-authentication · DNA encryption

A. Jayapalan (✉) · P. Praveenkumar · P. Savarinathan · T. Karuppasamy
SASTRA Deemed to be University, Thanjavur 613401, Tamil Nadu, India

© The Author(s), under exclusive license to Springer Nature Switzerland AG 2022
A. A. Abd El-Latif et al. (eds.), *Security and Privacy Preserving for IoT and 5G Networks*,
Studies in Big Data 95, https://doi.org/10.1007/978-3-030-85428-7_3

1 Introduction

Conventional wireless network has less adaptability when compared to the CR networks as spectrum sensing and subsequent utilization of the free spectrum do not occur efficiently. So, CR network is preferred as it is more versatile and flexible with dynamic spectrum access. The security aspect of the CR network is the notable issue. Not all the nodes can be accepted as the members of the network. Doing so will degrade the working performance of the network [1–3].

Cryptography is a method which transform the useful data into another format and could be retrieved back at the receiving end. In the literature, there many encryption methods have been proposed for secure end-to-end data transmission [4–13]. In this manuscript the message is encoded by a block cipher based cryptographic calculations [14, 15], the Rijndael Advanced Encryption Standard (AES) which is utilized by the Federal associations to secure sensitive information. Rijndael (articulated rain-dahl) is the calculation that has been chosen by the U.S. National Institute of Standards and Technology (NIST) as the possibility for the Advanced Encryption Standard (AES). Rijndael is a substitution linear transformation cipher, not requiring a Feistel organize [16]. Rijndael algorithm is used to encrypted the information transfer between the NM and CR and also between GH and CR.

In this innovative time, it is exceptionally basic to shield the computerized image information from extortion and forgery as they are transmitted over an open or public channel. Image encryption is the zenith of present-day cryptographic techniques. Image encryption assumes a significant job in the field of data covering up and furthermore gives a sheltered pathway to move of data through open systems, for example, the internet.

Now-a-days, DNA based cryptosystems appear to be the developing field for encrypting pictures. The principle focal points of DNA based calculations are gigantic parallelism, amazingly low power utilization, enormous information stockpiling and the idea of an unbreakable cryptosystem. In this paper, a hybrid encryption plot dependent on chaotic maps and Deoxyribo Nucleic Acid (DNA) [17, 18] is used which can be versatile for both specific and full image encryption.

2 Proposed Methodology

The proposed scheme involves bi- authentication process for a CR to become a member of node wherein it gains access to the network property using Network Manager and Group Head which make sure of user authorization. In first level, Network Manager interrogates the details about the CR. The interrogation is done by throwing questions to the CR. The question is encrypted and transmitted to the CR. The key to decode the question is known to the CR. The answer is encrypted and transmitted to the NM from CR. Based on the answers given by the CR the NM confirms about its authorization. If it confirms as an authorized user at the first level, the CR undergoes second level authentication where GH further interrogates details of user and reports to NM about the certified user so that it can access and enjoy the resources of the network. During authentication the questions and answers between the entities are encrypted and decrypted using Rijndael AES algorithm.

2.1   Rijndael AES Encryption

The Rijndael algorithm is symmetric, block cipher -based encryption technique that supports variable block size, key length and round number. It is capable of supporting different key lengths. They are 128, 192 and 256 bits. The rounds of iteration changes based on the key length. The Rijndael utilizes composite substitution and permutation techniques for the generation of keys.

Steps

Each round (with the exception of the last one) is a uniform and parallel structure of 4 stages.

- Sub Bytes (byte substitution utilizing an S-box)
- Shift Rows (a stage, which consistently moves the last three lines in the State)
- Mix Columns (substitution that utilizations Galois Fields, corps de Galois, $GF(2^8)$ number-crunching)
- Add Round key (bitwise XOR with an extended key) [19].

2.1.1   Key Generation

The initialization key vector required for encryption and decryption of the Rijndael AES is generated using the Sanskrit sutra mentioned as follows: The pseudo random sequence can be generated using Sanskrit sutra different structures during the substitution, scrambling and cyclic moving advances. The sutra contains eight syllables of both laghu and guru, where laghu and guru are logically represented as '0' and '1' respectively.

Laghu bit '0': **ya; ja; na; sa**
Guru bit '1': mā; tā; rā; bhā

The eight syllables of the sutra **ya mātārājabhānasa** form a memory wheel. Each syllable can be represented through cyclic rotation, thus forming 8 bit binary data for every seed syllable. It is shown in Fig. 1. This sutra can be used to generate pseudo random sequence [20].

Example of key generation

The key generated using sutras is
01101001110100011010001101000110100011110001110000011101001110100
110100111010001101000110100011010001111000111000011101001110100 11010   and
its hexadecimal representation is 74E8D1A3478E1D3A74E8D1A3478E1D3A.

Fig. 1 Sutras wheel

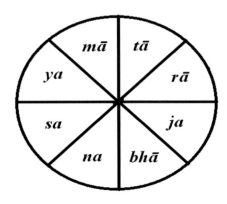

2.2 Bi-Authentication Scheme

Suppose if the CR wants to join the network, it has to send the request to the network manager. The network manager checks the whether the CR is trustworthy and reliable. To ensure it, the NM poses many questions to the CR. If all the questions are answered by the CR in a satisfactory manner, then only it is accepted to be member of the group. Due to malicious threat the questions asked by NM/GH are encrypted using Rijndael algorithm with the key generated using sutras cycle. The encrypted question is then sent to the CR through Orthogonal Frequency Division Multiplexing (OFDM) transceiver. The purpose of choosing OFDM transceiver is its ability to withstand severe noisy channel conditions. The reply from the CR is once again encrypted and sent to NM/GH through OFDM transceiver. Table 1 shows the sample questions to be asked by the NM and GH. The NM collects the basic information like ID and current location of the CR through the answers given by the CR to the NM. The information given by the CR is cross verified in the database by the NM. If the CR passes the test, then the NM asks the CR about the group which it wishes to join. The CR chooses the group near to its current location. The NM then conveys the message to the corresponding group head. The GH now enquires the CR by shooting it with remaining questions. If the group head is satisfactory about the answers given by the CR then it intimates to the NM. The NM now joins the CR as member of the group.

Table 1 Sample questions for authentication purpose

Questions		Asked by
Q1	What is your ID?	NM & GH
Q2	What is your current location?	NM & GH
Q3	In which group you intend to join?	NM & GH
Q4	What is your trust level?	GH
Q5	What is the joining code of the group?	GH

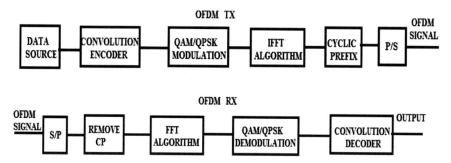

Fig. 2 OFDM transceiver

The block of Orthogonal Frequency Division Multiplexing (OFDM) based transmitter and receiver is as shown in Fig. 2.

The question to be asked which is listed in Table 1 is encrypted and then fed to the channel coder. Convolutional code with rate of ½ is used. Rate ½ means for every input given to the convolutional encoder two bits are obtained as output. Error control codes are meant to mitigate the channel noise. Then it is passed through mapper block. The modulation scheme chosen is 64 Quadrature Amplitude Modulation (QAM). 64 QAM means 6 bits make one symbol. This modulation scheme supports higher data rate. The output of signal mapper is converted into parallel data and then passed to IFFT block with the inclusion of Cyclic Prefix (CP). 128-point FFT is chosen. CP is meant to mitigate Inter Symbol Interference (ISI). 12% of the OFDM symbols are taken and fed as input as cyclic prefix. It is up sampled and then transmitted to the CR. The CR decodes the data and the answers to the question asked by the NM is once again encrypted and transmitted to the NM. The data is received and then CP is removed. It is then passed through FFT block. Its output is demodulated and channel decoding is performed with the aid of Viterbi decoder. The same process is repeated for the communication between GH and CR [21, 22].

It is assumed that the communication between the NM and the cognitive radio takes places in the Industry, Scientific and Medical (ISM) band. Since the data is encrypted, there is no issue in utilizing the ISM band. Additive White Gaussian Noise (AWGN) is considered as the channel noise.

2.3 DNA Algorithm

After becoming the member of the CR network, the CR utilizes the free channel and transmit the data to its intended receiver. To safeguard the information from malicious attack it is encrypted using DNA algorithm. Image encryption with DNA is superior to anything bit-level and pixel-level encryption plots as far as time and proficiency. A DNA arrangement involves four nucleic acid bases specifically, A (adenine), C (cytosine), G (guanine) and T (thymine). In a parallel framework 0 and 1 are integral.

Similarly, 00 and 11, 01 and 10 are complement to one another. Subsequently by utilizing these four DNA bases, one can plan 24 sorts of encoding rule. Notwithstanding, out of those 24 standards, just eight coding rules fulfill Watson–Crick supplement rule.

The RGB input image is spilt into Red, Green and Blue shaded planes. All the accompanying phases of encryption are done for each plane separately. Every pixel of the image is encoded utilizing a DNA rule. Chaotic sequences are created utilizing one dimensional logistic equations. Based on the chaotic sequences generated DNA encoding is performed to get the cipher picture. Decryption of the encoded data is done in the comparing request and the outcomes are investigated for the quality of encryption and other comparative parameters.

2.4   Pseucode Code

The following steps outline the process of the verification and the authorization of an incoming CR node to the established CR network.

Legend:

- CRN: Cognitive Radio Node, the node to be added into the network.
- NM: Network Manager, responsible for network management and monitoring.
- GH: Group Head, a sub unit of the NM, a portion of whose responsibilities are delegated to GH to reduce operational burden on the NM.
- MU: Malicious User, whose intensions are to disrupt the operations of the CR
- C: Channel
- E_(type)(C): Encrypted channel with type encryption
- CRN_status ( ): Current status of the Cognitive radio node

1. CR Node request initialisation
   a. CR node send RQT (request) to NM to join network
   b. NM sends AWK (acknowledgement) and proceeds to open an encrypted channel between CR and NM
   c. E_AES(C) is established.
   d. CRN_status(pending)

2. Verification by NM:
   a. Verification process is initiated by the NM. This is done by raising questions.
   b. Details of Node ID, Current Location, and Intent are recorded and verified against a database.
   c. if (details (CRN)) == details (NM database):
          verification control transferred to GH (next: step 3)
      else:
          connection RQT rejected (next: step 1)
   a. CRN_status(pending)
3. Verification by GH:
   a. Verification process is initiated by the GH. This is done by raising questions.
   b. Details of Node ID, Current Location, Internet, trust level and joining code are verified against data base.
   c. if (details (CRN)) == details (GH database):
          CRN is authorised and control transferred to NM (next: step 4)
      else:
          connection RQT rejected (next: step 1)
   d. CRN_status(pending)
4. Authorisation by GH:
   a. CRN is authorised by NM and appended into trusted list
      i. NM.trusted_list.append(CRN)
   b. Resources and channel BW are allocated to CRN
   c. CRN now may initiate contact.
   d. CRN_status(authorised)
5. Communication by CRN:
   a. CRN opens a encrypted channel to transmit sensitive images
      i. E_(DNA)(C) is established
   b. CRN is now able to exchange encrypted information between the authorised peers in NM.trusted_list.

3 Results and Discussion

Figure 3a–c shows the question asked by the NM, question encrypted using Rijndael AES algorithm and decrypted at the receiving end by the CR. The sample question take is here is "what is your id?". This question is encrypted using Rijndael algorithm with the key generated using sutras wheel. The key size is 128bit. Figure 3b shows the question in encrypted form. It is transmitted to the CR through OFDM transmitted with AWGN as channel noise. The CR receives the encrypted question through OFDM receiver and decrypts the information which is given in Fig. 3c.

Figure 4a, b shows the data transmitted through the OFDM transmitter and the data received utilizing OFDM based receiver. Additive White Gaussian Noise (AWGN) is the noise source added in the channel. Due to the presence of convolutional codes the noise has been properly mitigated. The likeness of the figures proves that the data has been transmitted and received correctly.

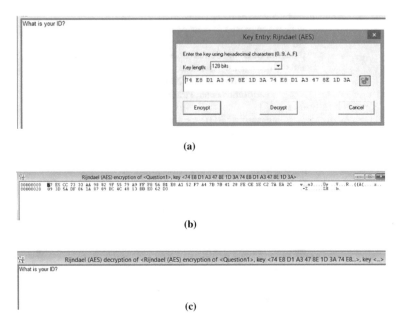

Fig. 3 **a** Sample question. **b** Encrypted form. **c** Decrypted form

Figure 5a–c shows the answer send by the CR to the NM, its encrypted form using the same algorithm and the decrypted one. Once the CR receives the question it answers the question. Figure 5a shows the answer (id number) sent by the CR to NM. 128 bit key is used to encrypt the id. Figure 5b shows the answer in encrypted form and it is encrypted using Rijndeal AES algorithm. The answer is transmitted to the CR using OFDM transmitted. AWGN is the channel noise. At the receiving end, NM receives the answer using OFDM receiver and decodes the answer which is shown in Fig. 5c.

Once the CR becomes the member of the network it is allocated with the free channel. Utilizing the free channel, it transmits the image to the intended receiver. Here a face of the man is chosen for transmission purpose. Figure 6a shows the original image to be transmitted. Figure 6b shows the image which is encrypted using DNA algorithm. DNA rules are followed for encryption and transmitted via AWGN channel. In the channel the image is corrupted with noise which is shown in Fig. 6c. The noise is removed at the received with the aid of Viterbi decoder. Once the noise is removed the image is decrypted to get back the original image and it is shown in Fig. 6d.

Figure 7 shows the sample database which exist with the network manager. The details given by the CR are cross verified with the details available in this database. If a match is found and it is satisfactory then only CR is accepted as member of the network.

Fig. 4 a OFDM
transmission. **b** OFDM
reception

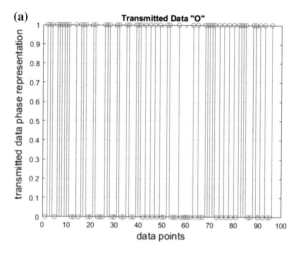

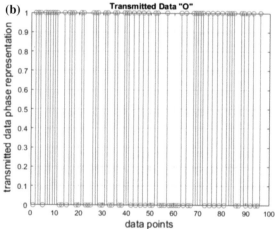

Validation

The degree of randomness of gray levels in the cipher image is accurately char-acterized by information entropy. Larger the entropy values imply the uniform dissemination of grayscale values. The entropy can be calculated as

$$\text{Entropy}(m) = \sum_{j=0}^{N-1} P(m_j) \log_2 \frac{1}{P(m_j)}$$

Here $N = 2^K$ is the total number of gray levels. The information entropy value for an ideal random image is 8. The achieved value of information entropy illustrates that it is tough to manipulate a successful attack.

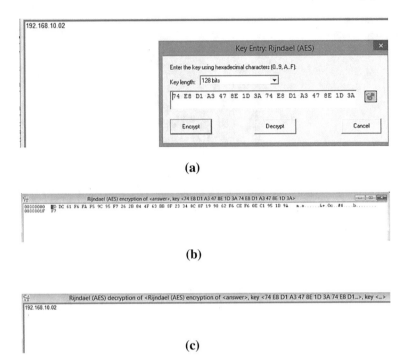

Fig. 5 **a** Answer sent by the CR. Answer in encrypted form. **b** Decrypted form

There are two typical methods to assess the quality of images. One is MSE (Mean Square Error) and the other is PSNR (Peak Signal to Noise Ratio). Using these two methods the difference between the plain-image and the cipher-image is determined.

The amount of similarity is calculated by taking the difference between plain image and cipher image and then computing the average of the error signal. Good quality of image have low MSE. The MSE is defined as

$$\text{MSE} = \frac{1}{MN} \sum_{i=1}^{M} \sum_{j=1}^{N} [I(i, j) - C(i, j)]^2$$

where.
   $I(i, j)$ -the no of pixels in plain image.
   $C(i, j)$ -the no of pixels in cipher image.
   i and j -the pixel position of the $M \times N$ image.

$$\text{MSE} = 0 \text{ when } I(i, j) = C(i, j)$$

The PSNR is inversely proportional the Mean Squared Error It is calculated in decibels. Higher the PSNR value better is the quality of the image. PNSR is defined

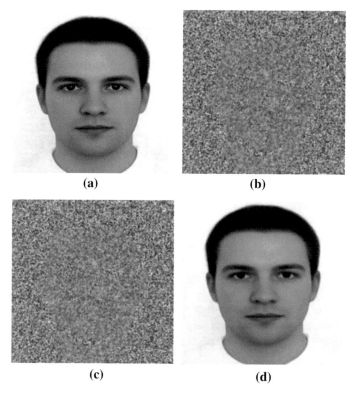

Fig. 6 **a** Original image. **b** Encrypted image. **c** Noise added image. **d** Decrypted image

Fig. 7 Sample database

Table 2 MSR, PSNR and entropy values

Image	MSE	PSNR(dB)	Entropy	
			Plain image	Cipher image
Face1	1.2084e + 04	7.3089	7.2788	7.9153
Pepper	7.8056e + 03	9.2067	7.4117	7.8402
panda	1.2321e + 04	7.2242	7.5914	7.9538
Flower	1.4291e + 04	6.5801	7.4199	7.9339
Lena	7.2893e + 03	9.5039	7.7758	7.7429
Tree	1.6245e + 04	6.0235	4.9319	7.9545

as

$$\text{PSNR} = 10 \times \log(\frac{P^2}{\sqrt{\text{MSE}}})$$

where,
    P-Peak signal value of plain image

$$P = \max(I(I, j), C(i, j))$$

Table 2 gives the values of MSE and PSNR for different images of size 256×256 respectively along with the entropy value. From the table it is clear that the information entropy for the encrypted images are approximately equal to a theoretical value of 8.

4 Conclusion

This work ensures to accept a trustworthy CR to become member of cognitive radio network. Integrity check has been carried to filter malicious nodes from useful node. To achieve this a two-level check has been proposed. At the initial screening the NM or the master controller enquires the CR by asking many questions. The questions are encrypted using Rijndael algorithm and transmitted. In the same way the answer from the CR is encrypted and transmitted to the NM. This encryption process ensures secure communication between the two parties. Once the NM is satisfied with the answers provided by CR it diverts it to GH. The CR undergoes second level of screening by answering the questions asked by the GH. On passing both the levels it now becomes member of the group. With the allocation of free channel the CR transmits the data to its intended receiver.

References

1. Khasawneh, M., Agarwal, A.: A secure and efficient authentication mechanism applied to cognitive radio networks. IEEE Access **5**, 15597–15608 (2017)
2. Truong, T., Nguyen, M., Kundu, C., Nguyen, L.D.: Secure cognitive radio networks with source selection and unreliable backhaul connections, IET Commun. **1215**, 1771–1777 (2018)
3. Jayapalan, A., Savarinathan, P., Praveenkumar, P., Karuppasamy, T.: Detecting and mitigating selfish primary users in cognitive radio. Wirel. Pers. Commun. **109**, 1021–1031(2019)
4. Zhang, T., Abd El-Latif, A.A., Han, Q., Niu, X.: Selective encryption for cartoon images. In: 2013 5th International Conference on Intelligent Human-Machine Systems and Cybernetics, vol. 2, pp. 198–201. IEEE (2013)
5. Li, L., El-Latif, A.A.A., Han, Q., Niu, X.: An improved additively homomorphic image encryption scheme based on elliptic curve Elgamal. Int. J. Adv. Comput. Technol. **4**(7), 223–230 (2012)
6. Abd-El-Atty, B., Iliyasu, A.M., Alanezi, A., Abd El-latif, A.A.: Optical image encryption based on quantum walks. Opt. Lasers Eng. **138**, 106403 (2021)
7. Beheri, M.H., Amin, M., Song, X., Abd El-Latif, A.A.: Quantum image encryption based on scrambling-diffusion (SD) approach. In: 2016 2nd International Conference on Frontiers of Signal Processing (ICFSP), pp. 43–47. IEEE (2016)
8. Abd El-Latif, A.A., Wang, N., Peng, J.-L., Li, Q., Niu, X.: A new encryption scheme for color images based on quantum chaotic system in transform domain. In: Fifth International Conference on Digital Image Processing (ICDIP 2013), vol. 8878, p. 88781S. International Society for Optics and Photonics (2013)
9. Zhang, T.J., Manhrawy, I.M., Abdo, A.A., Abd El-Latif, A.A., Rhouma, R.: Cryptanalysis of elementary cellular automata based image encryption. In: Advanced Materials Research, vol. 981, pp. 372–375. Trans Tech Publications Ltd, (2014)
10. Zaghloul, A., Zhang, T., Hou, H., M., Amin, Abd El-Latif, A.A., Abd El-Wahab, M.S.: A block encryption scheme for secure still visual data based on one-way coupled map lattice. Int. J. Secur. Appl. **8**(4), 89–100 (2014)
11. Abd El-Latif, A.A., Niu, X., Wang, N.: Chaotic image encryption using Bezier Curve in DCT domain scrambling. In: International Conference on Digital Enterprise and Information Systems, pp. 30–41. Springer, Berlin, Heidelberg (2011)
12. Mohamed, N.A., El-Azeim, M.A., Zaghloul, A., Abd El-Latif, A.A.: Image encryption scheme for secure digital images based on 3D cat map and turing machine. In: 2015 7th International Conference of Soft Computing and Pattern Recognition (SoCPaR), pp. 230–234. IEEE (2015)
13. Zhang, T.J., El-Latif, A.A.A., Amin, M., Zaghloul, A.: Diffusion-substitution mechanism for color image encryption based on multiple chaotic systems. In: Advanced Materials Research, vol. 981, pp. 327–330. Trans Tech Publications Ltd, (2014)
14. Kumar, A., Jakhar, S., Maakar, S.: Distinction between secret key and public key cryptography with existing glitches. Indian J. Educ. Inf. Manag. **19**, 392–395 (2012)
15. Preeti, S., Shende, P.: Symmetric key cryptography: current trends. Int. J. Comput. Sci. Mob. Comput. **312**, 410–415 (2014)
16. Wali, M.F., Rehan, M.: Effective coding and performance evaluation of the Rijndael algorithm (AES). In: Student Conference on Engineering Sciences and Technology, pp. 1–7 (2005)
17. Kaundal, A.K., Verma, A.K.: DNA based cryptography: a review. Int. J. Inf. Commun. Technol. **47**, 693–698 (2014)
18. Rahmana, N.H.U., Balamurugan, C., Mariappan, R.: A novel DNA computing based encryption and decryption algorithm. In: International Conference on Information and Communication Technologies, pp. 463–475 (2014)
19. Liu, N., Cai, J., Zeng, X., Lin, G., Chen, J.: Cryptographic performance for Rijndael and RC6 block ciphers. In: 11th IEEE International Conference on Anti-counterfeiting, Security, and Identification (ASID*)*, pp. 36–39 (2017)

20. Rajagopalan, S., Sharma, S., Arumugha, S., Upadhyay, H.N., Rayappan, J.B.B., Amirtharajan, R.: YRBS coding with logistic map-a novel sanskrit aphorism and chaos for image encryption, **788**, 10513–10541 (2018)
21. Anuradha, N.K.: BER analysis of conventional and wavelet based OFDM in LTE using different modulation techniques. Recent Adv. Eng. Comput. Sci. (RAECS), 1–4 (2014)
22. Manavi, F., Shayan, Y.R.: Implementation of an OFDM modem for the physical layer of IEEE 802.lla standard based on Xilinx Virtex-II FPGA. In: 59th Vehicular Technology Conference, pp. 1768–1772 (2004)

A Multi-fusion IoT Authentication System Based on Internal Deep Fusion of ECG Signals

Basma Abd El-Rahiem and Mohamed Hammad

Abstract Recently, the interest in using wearable devices or the internet of things (IoT)-based biometric authentication, especially IoT-based electrocardiogram (ECG) has increased. ECG-based biometric authentication has received great attention as a next-generation promising technique and been implemented with various approaches to improve the authentication performance for the past few decades. However, ECG signals of a person may vary according to his/her physical states, or health conditions, possibly leading to authentication failure in some cases. Therefore, it is essential to design a robust method that handles the ECG subject variability for accurate authentication. In this Chapter, we proposed an efficient and robust authentication system based on ECG. In this study, we propose a novel deep learning fusion framework using the transfer learning concept where the deep features extracted from different models are combined into a single feature which are then fed to a custom classifier such as a support vector machine (SVM) for authentication. Cross-validation studies are used to assess the performance of the proposed authentication system using two public databases. Evaluation results show that the performance of our fusion model achieved an authentication accuracy of 99.4% with a high level of precision and recall. Finally, the results show that the proposed system is suitable for real-time applications.

Keywords Authentication · Biometric · Deep learning · ECG · Fusion · IoT · Transfer learning · SVM

B. Abd El-Rahiem
Mathematics and Computer Science Department, Faculty of Science, Menoufia University, Shebin El-Koom, Egypt

M. Hammad (✉)
Information Technology Department, Faculty of Computers and Information, Menoufia University, Menoufia, Egypt
e-mail: mohammed.adel@ci.menofia.edu.eg

© The Author(s), under exclusive license to Springer Nature Switzerland AG 2022
A. A. Abd El-Latif et al. (eds.), *Security and Privacy Preserving for IoT and 5G Networks*, Studies in Big Data 95, https://doi.org/10.1007/978-3-030-85428-7_4

1 Introduction

Biometric is the statistical analysis of the unique physical and behavioral characteristics of individuals. This technology is mainly used to identify and control access,
or to identify individuals who are under surveillance. The basic premise of biometric
authentication is that each individual can be accurately identified, by their characteristics. Physical or behavioral subjectivity, the term biometric is derived from
the Greek words bio meaning life, and metric meaning scale. The two main types of
biometric identifiers are based on physiological or behavioral characteristics. Physiological identifiers are related to the composition of the authenticated user and include
facial recognition [1–5], fingerprints [6–8], finger geometry (size and position of the
fingers) [9], iris recognition [10, 11], vein recognition [12], retina scanning [13],
voice recognition [14] and DNA matching [15] as shown in Fig. 1. While behavioral
identifiers include unique ways in which individuals act, including electrocardiogram
(ECG) [16], electroencephalogram (EEG) [17], recognition of writing patterns [18],
gait [19], and some of these behavioral identifiers can be used to provide continuous
authentication rather than a one-time authentication check. Biometric properties can
be summarized as follows:

- **Unique**: It should be different from one person to another, even twin brothers.
- **Universality**: It must be a universal characteristic and not present in a certain
  class of people.
- **Permanent**: It should not be affected by age and be permanent.
- **Measurable**: It must be measurable with simple technical tools.
- **Easy to use**: It should be easy and comfortable to measure.

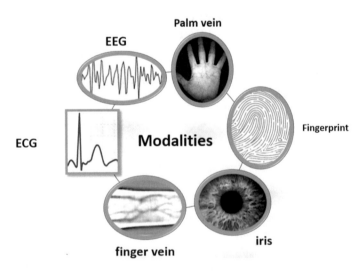

Fig. 1 Examples of physiological and behavioural biometric modalities

Authentication through biometric verification is becoming increasingly common in corporate and public security systems [20], consumer electronics [21], and point-of-sale applications [22]. In addition to security, the driving force behind biometric verification has been apt, as there are no passwords or remembering security codes to carry, and it can some biometric methods, such as measuring a person's gait, can operate without any direct contact with the person being monitored. Components of biometric devices include:

- A reader or scanning device to record a biological agent that is documented.
- A program for converting the scanned biometric data into a unified digital format and comparing points that match the observed data with the stored data.
- A database to securely store biometric data for comparison.
- Biometric data can be kept in a central database, although modern biometric applications mostly rely on collecting biometric data locally and then hashed it encrypted, so that documentation or identification can be achieved without direct access to the biometric data itself.

The biggest privacy issue with biometric use is that physical features such as fingerprints and retinal vascular patterns are generally fixed and cannot be modified, and this distinguishes between non-standard factors such as passwords (something you know) and unique tokens (something you have), which could be replaced if it was hacked, as the fingerprints of more than 20 million individuals were compromised in the 2014 U.S. Bureau of Staff. In 2015 Jan Chrysler, also known as "Starbug," a biostatistics researcher at Chaos Computer Club demonstrated a way to extract enough data from a high-resolution image to defeat iris scan authentication, and in 2017, Krissler reported defeating the scanner authentication system. The iris scanner used by the Samsung Galaxy S8 smartphone, Chrysler had previously recreated its user fingerprint from a high-resolution image, to demonstrate that Touch ID's fingerprint authentication system was also weak [23]. To overcome these limitations of physical biometrics, we proposed our algorithm based on one of the behavioral biometrics, which is ECG for human authentication.

The ECG is the electrical current that moves through the heart as it beats. The movement of an electrical current is divided into parts, and each part is given an alphabetical designation on the ECG. Each heartbeat begins with a pulse or signal from the heart pacemaker (the sinoatrial node or sinus node). This signal stimulates the two upper chambers of the heart (atria). The P wave indicates atria activation. Then, electrical current flows into the lower chambers of the heart (ventricles). The QRS complex represents the activation of the ventricles in the ECG. The electric current then spreads back to the ventricles in the opposite direction. This activity is called the recovery wave, which is represented by the T wave as shown in Fig. 2. Recently, several studies have been used ECG signals for identification and authentication [24–43] as the characteristics of ECG signals have the following advantages:

- University: this requirement is satisfied for ECG signals as these signals can be monitored from every person.

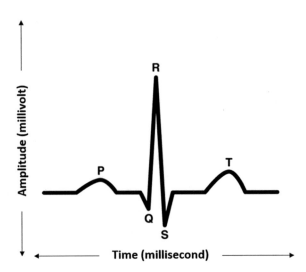

- Permanence: this requirement is also satisfied for ECG signals as the structure of these signals is invariant over time.
- Robust: the ECG systems are naturally more robust to spoof attacks comparing with other biometrics such as a fingerprint.
- Liveness Detection: ECG is only present in a living subject, which can ensure sensor liveness unlike other biometrics such as fingerprint which requires additional processing to establish the liveness.

The main goals of this Chapter are as follow:

- The first goal is to propose a new ECG biometric authentication system based on deep transfer learning that is more robust and efficient than other previous authentication systems.
- The second goal is working on large population size, where most of the previous studies worked on small data which not suitable for real life applications.
- The third goal is to explore the effectiveness of using internal fusion with deep learning models, which improve the performance of ECG biometric systems.

The main contributions of this Chapter are the following:

1. Proposed a novel ECG authentication system based on deep transfer learning. The novelty of the proposed system is in the feature extraction stage, where we combine the output features from several pre-train models.
2. Proposed a new internal fusion to combine the internal feature vectors for each pre-train model, which increases the performance of the whole system.
3. The performance of the proposed authentication system is evaluated using ECG records from a big data such as MWM-HIT database [41] under different conditions, which make our system is suitable for real time authentication.
4. An efficient feature extraction method has been employed to extract the deep features from the ECG signals to obtain better authentication performance.

5. Proposed a new, efficient and robust internet of things (IoT) authentication system based on ECG signals.

This Chapter is organized as follows: Second Section discuss briefly the previous machine learning and deep learning studied based on ECG for human authentication. After that, in the third Section, the effect of ECG in IoT system is discussed. In the fourth Section, the proposed method based on ECG for authentication is presented. Results are discussed in Section five and the conclusion of this Chapter is discussed in the last Section.

2 Literature Review

In this Section, several works using machine and deep learning methods related to ECG biometric for human authentication are discussed. Recently, several previous researchers worked on different biometrics for human authentication [24–43]. For authentication based on ECG, many previous works used machine learning for human authentication based on ECG [24–33], however, very few studies used deep learning for human authentication based on ECG [34–43]. In this Section, we divided the previous studies into two categories, in the first category, we focused on the studies based on machine learning methods, where these studies perform all stages of machine learning methods such as preprocessing, feature extraction and classification. While the second category focused on the studies-based deep learning methods, where the common deep model that used in these studies is the convolutional neural network (CNN) model.

2.1 Previous Studies Based on Machine Learning Methods

Recently, using ECG as a biometric for authentication is increased and several works based on machine learning methods have attempted to use ECG as a single biometric as follows.

Altan et al. [24], used the ECG signals for human identification based on Second Order Difference Plot (SODP) as a feature extraction method and k-Nearest Neighbor (KNN) algorithm as a classifier for classification. They worked on three ECG databases named ECG-ID database (ECG-ID) [44, 45], MIT-BIH Arrhythmia (ADB) [45, 46] and MIT-BIH Normal Sinus Rhythm (NSRDB) database [45] for identification. They obtained the highest accuracy of 99.86% using the ADB.

Goshvarpour and Goshvarpour [25], developed a system based on a sparse algorithm using a non-fiducial one-lead ECG feature set for identification. They used a matching pursuit (MP) for extracting the features and probabilistic neural network (PNN) and KNN for classification. The features are ranked and selected using principal component analysis (PCA) and linear discriminant analysis (LDA). They achieved the highest recognition rate of 99.68% using PNN as a classifier.

Pinto et al. [26], introduced a system for identification and authentication tasks based on ECG signals. They used Discrete Cosine Transform (DCT) and Haar transform for feature extraction and Support Vector Machines (SVM), KNN, Multilayer Perceptrons (MLP), and Gaussian Mixture Models for classification. They obtained a 94.9% identification rate (IDR) and a 2.66% authentication equal error rate (EER).

Alotaiby et al. [27], presented an identification system based on non-fiducial ECG signals. They presented the common spatial pattern (CSP) method for feature extraction and SVM for classification. They obtained the best identification rate of 98.92% with an EER of 0.08 using a single chest-based lead from 200 subjects of the Physikalisch-Technische Bundesanstalt (PTB) database [45, 47].

El_Rahman [28], presented a human recognition system based on ECG signals. The author used Pan and Tompkins [48] approach for extracting the ECG features and NN, Fuzzy Logic (FL) and Nearest Mean Classifier (NMC) for classification. The method worked on the MIT-BIH database and obtained an accuracy of 98.99%.

Barros et al. [29], presented a data improvement model for ECG identification. They used local maximum points as a feature from the input ECG signals and used random forest (RF) for classification. They used data augmentation techniques to increase the number of data used to be suitable for real applications. They obtained mean false acceptance rates of 0.0194%, mean false rejection rates of 38.38%.

Goshvarpour and Goshvarpour [30], presented a human identification system based on ECG signals. They used Cauchy–Schwartz divergence (CSD), Euclidean distance (ED), Cauchy–Schwartz quadratic mutual information (CSQMI), Euclidean distance quadratic mutual information (EDQMI), and cross information potential (CIP) for extracting the features and KNN for classification. They obtained the highest average accuracy rate of 97.62%.

Wang et al. [31], presented a framework based on ECG for human authentication. They used a Multi-Scale Differential Feature (MSDF) and Collective Matrix Factorization (CMF) for feature extraction and the Euclidean distance for the matching step. They obtained an EER of 0.0255 using the PTB database.

Liu et al. [32], introduced an identification method based on ECG signals. They used multiscale autoregressive model (MSARM) to extract the ECG features. Finally, they developed a random forest-based authentication system for classification. They obtained a recognition rate of 93.15%.

Hammad et al. [33], presented an authentication system based on ECG signals. They used Pan-Tompkins algorithm to extract the ECG features and used two cancelable biometric methods to protect these features and finally, they used Feed-Forward Neural Network (FFNN) as a classifier for authentication. They obtained an EER of 0.14 using PTB database. Table 1 summarizes the previous works that used machine learning methods for ECG authentication.

Table 1 Summary of state of art algorithms based on machine learning

Authors/Year	Methodology	Database	Performance
Altan et al. [24]	SODP KNN	ECG-ID MIT-BIH NSRDB	Accuracy 91.96% 99.86% 95.12%
Goshvarpour and Goshvarpour [25]	MP PNN KNN PCA LDA	ECG-ID	Accuracy: 99.04%
Alotaiby et al. [27]	CSP SVM	PTB	Identification rate: 95.15% EER: 0.1
A. El_Rahman [28]	NN FL NMC	MIT-BIH	Accuracy: 98.99%
Wang et al. [31]	MSDF CMF	PTB	EER: 0.0255
Liu et al. [32]	MSARM Random forest	SIAT-ECG database	TRR: 98.99% TAR: 95.04%
Hammad et al. [33]	FFNN Cancelable algorithms	PTB	EER: 0.14

2.2 Previous Studies Based on Deep Learning Methods

Recently, using ECG as a biometric for authentication is increased and several works based on deep learning methods have attempted to use ECG as a single biometric as follows.

Zhao et al. [34], presented an authentication method based on ECG signals using S-transformation and CNN. They first converted the one-dimensional ECG signal to a two-dimensional image and fed it to the CNN. After that, they used CNN for feature extraction and classification. They obtained the best accuracy of 96.63% and EER of 5.68%.

Chu et al. [35], introduced an authentication system based on ECG using the parallel multiscale one-dimensional residual network. They used the parallel network for extracting the features from the ECG vector. They used Adam optimizer to train the network faster and separated the classifier for classification. They achieved an accuracy of 98.24% and EER of 2%.

Labati et al. [36], presented an ECG recognition method based on deep CNN. They used the presented deep learning model for feature extraction and classification. They also used data augmentation techniques to enlarge the number of training data. They obtained the best EER of 2.15%.

Abdeldayem and Bourlai [37] presented an ECG-based identification system using spectral correlation and CNN. They used autocorrelation function (ACF) and spectral correlation function (SCF) for feature extraction and CNN for classification. They obtained an accuracy of 95.6% and a false acceptance rate of 0.2%.

Kim and Pyun [38] presented a real-time identification system based on ECG. They used bidirectional long short-term memory (LSTM)-based deep recurrent neural networks (DRNN) through late-fusion for feature extraction and classification. They obtained an overall accuracy of 99.73%.

Kim and Pan [39] presented a real-time ECG recognition system using deep learning based on 1-D ensemble networks. They used the ensemble networks for feature extraction and classification. They achieved an average recognition accuracy rate of 99.8%.

Hammad et al. [40], presented human authentication system based on ECG using ResNet-Attention model. They introduced two end-to-end models; the first model using CNN and the second model using ResNet with an attention mechanism called ResNet-Attention and used these models for feature extraction and classification. They achieved the best accuracy of 99.27%.

Hammad et al. [41], presented an ECG authentication system using CNN. They used scanning and removing methods for extracting the features and used CNN for classification. They also introduced a new database called MWM-HIT database and obtained the highest accuracy of 99.30% and EER of 1.63%.

Hammad and Wang [42] presented a human authentication system based on ECG. They extracted the features using CNN model and used the cancelable technique (matrix operation) to protect these features. Finally, they presented Q-Gaussian multi-class support vector machine (QG- MSVM) [8] for classification. They achieved an overall accuracy of 96.56%.

Hammad et al. [43], also introduced an authentication system based on ECG using deep CNN model and QG-MSVM. They used CNN for extracting the deep features of the input ECG image and then protect these features using cancelable technique (improved Bio-Hashing) and finally perform the classification using QG-MSVM classifier. They obtained the highest accuracy of 98.97%. Table 2 summarizes the previous works that used deep learning methods for ECG authentication.

The previous machine and deep learning methods have many limitations such as:

- Most machine learning methods suffer from over fitting, which can affect the results in a different database.
- Several works need big data to obtain high accuracy and obtained very low accuracy on small data.
- Several works worked on a complex system with the high cost and low speed.
- There is no guarantee that there is liveness detection in most of the previous works.

In this Chapter, we overcome these limitations by proposing an efficient and robust method based on pre-trained deep models with low cost and high accuracy for human authentication.

Table 2 Summary of state of art algorithms based on deep learning

Authors/Year	Methodology	Database	Performance
Zhao et al. [34]	S-transformation CNN	ECG-ID	Accuracy: 96.63% EER: 5.68%
Chu et al. [35]	parallel multiscale one-dimensional residual network	ECG-ID PTB MIT-BIH	EER: 2.00% 0.59% 4.74%
Kim and Pyun [38]	LSTM DRNN	NSRDB MIT-BIH	Accuracy: 100% 99.8%
Hammad et al. [40]	ResNet-Attention CNN	PTB CYBHi	Accuracy: 98.85 99.27%
Hammad et al. [41]	CNN	MWM-HIT	Accuracy: 99.3% EER: 1.63%
Hammad and Wang [42]	CNN QG-MSVM	PTB	Accuracy: 96.56%
Hammad et al. [43]	CNN QG-MSVM	PTB CYBHi	Accuracy: 98.66% 98.97%

3 IoT Authentication System

In this Section, we discussed how the proposed system can connect with IoT devices and discussed the new technologies of using ECG signals for authentication. IoT has been one of the hottest topics in technology in recent years, as many IoT devices are reaching the industrial, consumer, and commercial markets, from smart devices, TVs and thermostats to authentication systems, medical devices, and business tools [49–52]. Now more devices than ever before are able to communicate and can be controlled over the Internet.

Recently, wearable devices (e.g. ECG devices) enable recovering data from their users among supplementary applications. Several services can be devised when combining these devices with IoT. Several approaches based on IoT have been proposed to apply user data, and particularly ECG signals, for biometric authentication.

Several IoT applications were launched recently that used ECG for authentication such as the Nymi bracelet [53] which is shown in Fig. 3. The bracelet includes a sensor that measures the "ECG" that records the small electrical impulses issued by the heart. The sensor enables a user's heartbeat to be matched with a pre-stored pattern to verify his identity. With regard to the security aspect, it is not possible to obtain an ECG of the user without his consent, unlike the fingerprints that people leave behind samples of when touching things, which allows the possibility of copying or falsification. The ECG is internal, which makes it difficult to obtain the user's

Fig. 3 Nymi bracelet [53]

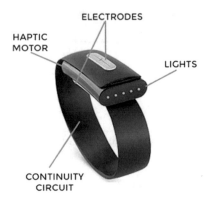

identity. As for the convenience feature, it comes from the need for a "Nymi" user to confirm his/her identity once a day, instead of having to provide a fingerprint before each transaction. Once the user's identity is verified, the user can permanently and reliably access several services and devices via a wireless connection. In addition, the ECG sensor can collect signals continuously until it reaches a reading identical to the stored model, which solves the problem of users' need to re-position their fingers on the fingerprint sensor in order to be able to read them correctly. To use the "Nymi" bracelet, once the user's identity is verified, the user can deal with his/her car and electronic devices, pay money and open the hotel room without the need to enter passwords. After registration, the band can be presented as certified RPs terminals. These terminals also operate NEA, although not provided for the Nymi Band. NEA Terminal and Nymi Band share bidirectional ECDSA signature verification [53]. The station's NEA retrieves an ad number by calling the Nymi API function and passing it to the RP server. Server signs Nymi Band hash with the nonce server. This can be any server-generated random value. This signature is sent to Nymi Band for verification. Besides signing, the nonce server and RP public key are also sent. The Nymi uses the RP public key to verify the signature as shown in Fig. 4. From the Figure, if the bracelet verifies the user successfully, it ciphers the server with its private key and sends to the server for verification. If the verification is successful, the server sent a notification to accept the bracelet for authentication and allow the user to access it.

4 Method

In this Section, we discussed in detail the proposed system based on ECG for authentication, which consists of 4 stages such as preprocessing, feature extraction, feature reduction and authentication stages. From the discussion in the literature study, we observe that the deep fusion techniques have not been fully leveraged for a robust ECG authentication system. To this end, we have proposed a deep learning based

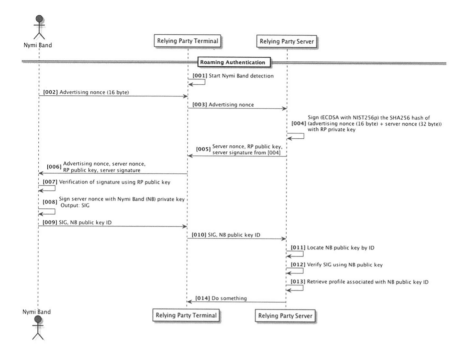

Fig. 4 Nymi bracelet as a roaming authentication [53]

the fusion model that extracts the deep features from the input ECG that are used to obtain a robust authentication of these signals. In this Section, we start by describing the various components of our proposed system and the internal fusion method.

Figure 5 shows the structure of the proposed system, which consists of the following stages:

- **Preprocessing stage**: In this stage, all ECG input signals are normalized, and the noise are removed using several filtering algorithms. The pre-processed image data is then split into training, validation, and test sets, from which we have used the training and validation data to train and validate our models through tenfold cross-validation [54].
- **Feature extraction stage**: In this stage, several deep pre-trained models are employed to extract the deep features of each ECG signal.
- **Feature reduction stage**: In this stage, the extracted features from the pre-trained models are reduced using several common reduction algorithms to prepare the features for the authentication stage.
- **Authentication stage**: In this stage, the reduction features are fed into several common classifiers for authentication, where the result of the classifier is the final decision of the system (accept or reject). The performance of the proposed system is measured with the test dataset using standard metrics.

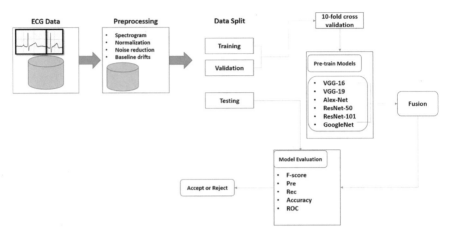

Fig. 5 Block diagram for the proposed authentication system

4.1 Preprocessing Stage

Robust preprocessing algorithms are needed to have high authentication accuracy. This preliminary step reduces the noise from the input ECG signals, smoothens the ECG signal and reduces drift suppression and baseline wander. This preprocessing step makes the ECG signal suitable for subsequent processes. The most common methods used to reduce signal noise are (i) Second-order low-pass Butterworth filtering [55], (ii) Daubechies wavelet 6 (db6) [56]; and (iii) Orthogonal wavelet filter [57]. In this Chapter, we worked on Second-order low-pass Butterworth filtering to reduce the noise of the input ECG signals. Then, *five*-order moving average filtering is used to make the signal smooth [58]. Finally, for reducing drift suppression and baseline wander, we used high pass filtering [59]. We worked on the whole ECG signals without the need for using any algorithms for segmentation, which is satisfied the real-world applications. Finally, we converted the 1-D ECG signals to 2-D ECG spectrogram with a specific size of 224×224. The result of the preprocessing stage (spectrogram ECG) is fed to the different pre-train models.

4.2 Feature Extraction Stage

The rapid developments in Computer Vision, and by extension—image classification has been further accelerated by the advent of Transfer Learning. To put it simply, transfer learning allows us to use a pre-existing model, trained on a huge dataset, for our own tasks. Consequently, reducing the cost of training new deep learning models and since the datasets have been vetted, we can be assured of the quality.

In this study, several pre-trained CNN models were employed for ECG authentication. Figure 6 shows these pre-trained which are called: VGG-16, VGG-19, Alex-Net, ResNet-50, ResNet-101 and GoogleNet.

VGG-16: This model [60] is one of the common pre-trained deep models, which was proposed by Simonyan et al. [61]. The model was trained on 14 million images with 1000 classes. As shown in Fig. 6, the input of this model is the ECG spectrogram of dimension 224 × 224 passed through five blocks of convolution (conv) layers with a total of 13 conv layers. The size of the filter that used in these conv layers is very small (3 × 3) with the same padding (1-pixel). Each conv block is followed by a max pooling (maxpool) layer of stride (2, 2). Finally, there are three fully connected (fc) layers passed to the Softmax layer for classification and the ReLU activation

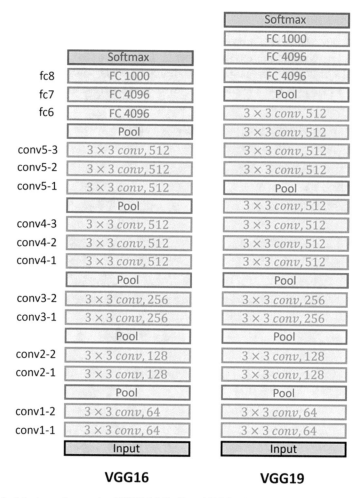

Fig. 6 Architecture of pre-trained VGG-16 (Left) and VGG-19 (Right) [61]

function [62] was used in all hidden layers. However, this model is very slow to train the images and it needs high size and bandwidth due to the depth of this model which makes it inefficient.

VGG-19: This model [61] is the same as VGG-16, however, it is deeper architecture than VGG-16 with more layers as shown in Fig. 6. The input of this model is the ECG spectrogram of dimension 224 × 224 passed through five blocks of conv layers with a total of 16 conv layers and three fc layers.

Alex-Net: This model is 8 layers deep and was trained on millions of images with 1000 classes. As shown in Fig. 7, the input of this model is the ECG spectrogram of dimension 224 × 224 passed through 5 conv layers and 3 fc layers. The ReLU activation function was applied after every layer and there are dropout layers [63], which were applied before the first and second fc layer. This network was trained for six days simultaneously on two GPUs.

ResNet: This model is based on residual learning, which can simplify the training of networks by considering the input layer as a reference [64] as shown in Fig. 8. This model is widely used among other pre-trained models, which has 48 conv layers along with 1 max pool and 1 average pool layer as in ResNet-50 and 101 layers deep with 33 residual blocks as in ResNet-101.

GoogleNet: is a CNN that proposed by research at Google with 22 layers deep [61]. This architecture was the winner at the ILSVRC 2014 image classification challenge.

Fig. 7 Architecture of pre-trained AlexNet [63]

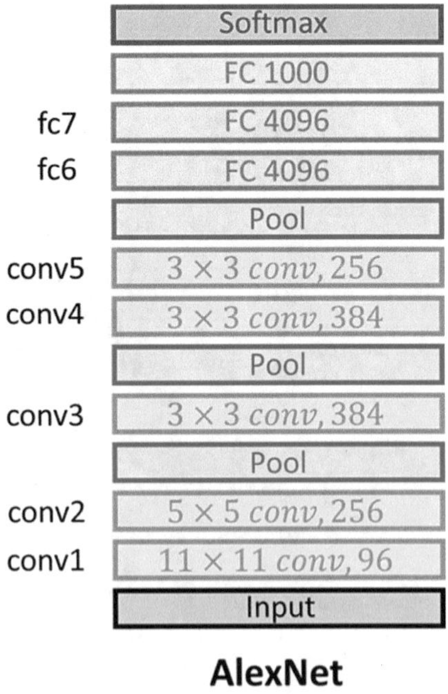

	Softmax
	FC 1000
fc7	FC 4096
fc6	FC 4096
	Pool
conv5	$3 \times 3\ conv, 256$
conv4	$3 \times 3\ conv, 384$
	Pool
conv3	$3 \times 3\ conv, 384$
	Pool
conv2	$5 \times 5\ conv, 256$
conv1	$11 \times 11\ conv, 96$
	Input

AlexNet

Fig. 8 Architecture of pre-trained ResNet [64]

It uses 1×1 convolution and global average pooling that enables it to create deeper architecture. This architecture takes the ECG spectrogram of dimension 224×224. All the convolutions inside this architecture use ReLU as their activation functions.

As part of fine-tuning, we extracted the features from the last fully connected (e.g. for VGG-16, VGG-19 and AlexNet the number of features are 1000) and then fed it to the classifier. We have deleted the classification part in each model and perform the classification using separate classifiers such as SVM or KNN.

4.3 Feature Reduction Stage and Internal Fusion

In this step, we select the features from the output of the fully connected layers in all models, which give the highest effect on the authentication accuracy comparing to other layers. We reduced the feature size of the selected features of each biometric by fusing all selected feature sets using canonical correlation analysis (CCA) that introduced by Haghighat et al. [65]. The main idea of CCA is to find the correlation between all selected feature sets. Figure 9 shows the process followed for internal feature fusion of ECG using CCA. After selecting the deep features from each model, a new transformation of these features is calculated based on CCA. Finally, we fuse these transformed features using addition or concatenation to represent each model by single features. Finally, fusing between all models using addition or concatenation fusion. In this fusion, we fuse two or more appropriate features extracted from the ECG images into one appropriate feature vector with additional intense information than the input features. After applying this fusion, the feature vector will comprise deep information to explain the ECG image properly. Therefore, this fusion considers being a robust and efficient approach. Figure 10 shows the overall system after fusion of all pre-trained models. We obtained the final features which fed to the suitable classifiers (SVM or KNN).

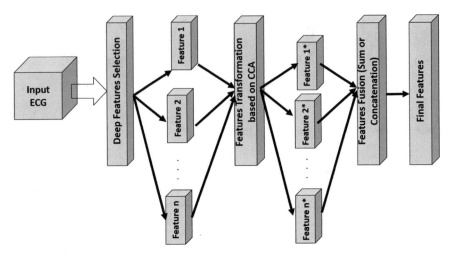

Fig. 9 Internal deep Fusion based on CCA

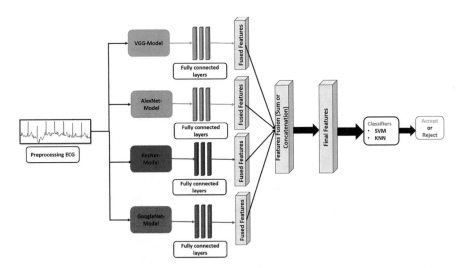

Fig. 10 Overall the authentication system after fusion the features of each model

4.4 Authentication Stage

In this step, we employed *five* well-known classifiers and selected the classifier that shows the best authentication accuracy. These classifiers are as follow:

a Support Vector Machine (SVM)

It is a supervised learning algorithm that can be employed for regression and classification problems by finding a hyperplane that splits the features into diverse domains [66, 67]. There are *two* main types of SVM, namely, Linear SVM

(LSVM) for *two*-class problems and Non-Linear SVM for multi-class problems. In this work, we employed LSVM and selected the regularization parameters via the *five*-fold cross-validation approach from 1^{-1} to 1^{+5}.

b   K-Nearest Neighbors (KNN)

It is one of the algorithms of machine learning within the supervised learning group, which is one of the simplest algorithms due to its ease of use and consumption for a little time. The KNN is sensitive to errors during the training sequence [68, 69]. In this work, we tested the KNN when $K = 1$–20.

c   Random Forest

Random Forest produces one of the most powerful and fully automated machine learning technologies. It is able to handle unbalanced data containing missing values, and this maintains accuracy when there is a large percentage of missing data. However, the size of the model may exceed the amount of data that is designed for analysis. Therefore, it is not suitable for analyzing the complex data structures included in data sets that potentially contain millions of columns with a limited number of rows [70].

d   Naïve Bays

Naive Bayes' theorem is one of the most popular methods of machine learning, data analysis, and classification characterization where it is done quickly in processing and efficient in forecasting operations. It has a higher speed for large training sets and works well for small-scale data. However, it is sensitive to noisy data and gives errors in the classification decision [71].

e   Artificial Neural Networks

Artificial neural networks consist of a set of algorithms that simulate an evolving human brain and manufacture electronic brains that are capable of learning and developing as well as the human brain. They are made up of nodes that look like nerve nodes in the human nervous system. The artificial neural network nodes are organized into parallel layers, the input, hidden and output layers. These layers are bound together to form the entire neural network [72].

5   Results and Discussion

To demonstrate the efficacy of our proposed system in authenticating the ECG signals, we extensively evaluate and compare the performance results of the model with fine-tuned transfer learning models, namely, VGG-16, VGG-19, Alex-Net, ResNet-50, ResNet-101 and GoogleNet using two publicly ECG datasets. The experiments have been performed in MATLAB 2019a version on Windows 2010 ultimate with Intel core i7 processor and 16 GB RAM. To evaluate the proposed system, we employed a tenfold cross-validation technique [54].

Table 3 Descriptions of the MWM-HIT database

Conditions	Sitting	Standing	Supine	Standing with exercise	Sitting with exercise
# of records	100	100	100	100	100
Age range	22–60 years old				
Gender	66% Male and 34% Female from Total subjects of 100				

Fig. 11 Samples of ECG images from MWM-HIT database

5.1 ECG Databases

In this Chapter, the MWM-HIT is employed [41]. This database contains 10 s ECG recordings from 100 subjects each subject has 5 different records with different conditions which are in total 500 records. The five conditions are sitting, standing, supine, exercise sitting and exercise standing. The Carewell ECG Workstation (PCECG-500) was used for recording. It includes four parts, the collection box, the ECG information management software PCECG-500 A, the accessory box, and a PC. The recordings were digitized at 1000 samples per second (1000 Hz). Four electrodes are used to capture the ECG from the subjects' body like in the leads I, II and III configurations (which are actually Willem Einthoven's original leads). The four electrodes relate to the left hand, right hand, left leg, and right leg.

Table 3 shows the characteristics of this database. Figure 11 shows samples of the MWM-HIT database.

In addition, PTB Diagnostic ECG [45, 47] is employed with MWM-HIT to evaluate the proposed authentication system and for the comparison with other previous methods. The PTB ECG database provides data from 290 subjects. Each file with one or more signals (1–7 signals per patient file), which contains the standard 12 ECG leads plus the X, Y, Z Frank leads. The ECG signals are digitized at 1000 samples per second (1000 Hz), with 16-bit resolution over a range of ± 16.384 mV.

5.2 Performance Metrics

The F-score metric was selected for evaluation of the proposed authentication system. F-score is the harmonic mean of precision (Pre) and recall (Rec). The Pre is also known as the confidence or positive predictive value and the Rec is otherwise called the sensitivity or true positive rate. Pre and Rec are both expressed mathematically as shown in (1) and (2). The higher the F-score, the better the method. F-score is given in (3).

$$Pre = \frac{TP}{TP + FP} \tag{1}$$

$$Rec = \frac{TP}{TP + FN} \tag{2}$$

$$F - score = 2\frac{Pre \times Rec}{Pre + Rec} \tag{3}$$

where TP, FP and FN are true positive, false positive and false negative respectively. A higher F-score correspond to better authentication. In addition, we used also Receiver Operating Characteristic Curve (ROC) as our performance metric.

5.3   Results of the Authentication System

The results obtained from testing the system are stated below. The stated results are two, the first is the f-score and average time of the ECG authentication algorithm. The second result states the FAR, FRR and Accuracy of the system. To analyze the experimental results of our system on the mentioned datasets, we employ several pre-trained models and apply them on ECG signals to extract the deep features. After that, we select the fully connected layers to describe these features from each model and make the internal fusion between the layers in each model. Then, we employ two fusion methods summation and concatenation methods to fuse the selected deep features from each model. For authentication, we have selected a linear SVM classifier, which has shown its effectiveness in the two databases. The summary of the parameters and functions used for model training is shown in Table 4.

The confusion matrices of the proposed system with SVM and KNN classifier on MWM-HIT database are shown in Figs. 12 and 13, respectively.

From the previous confusion matrices, we can observe that 0.6% of the imposter's records are wrongly authenticated as accepted users and 100% of the accepted users are correctly authenticated when using SVM classifier with the proposed fusion system. In addition, we can observe that 95% are correctly authenticated as imposters and 0.7% are wrongly authenticated as an accepted user using KNN classifier.

Table 4 Summary of parameters and functions used for training

Epoch numbers	100
Batch size	8
Optimizer	Adam
Loss function	Categorical

B. Abd El-Rahiem and M. Hammad

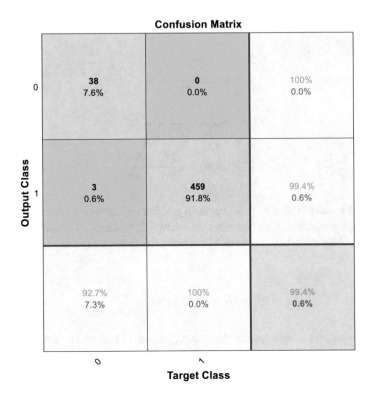

Fig. 12 Confusion Matrix of the proposed system using SVM classifier on MWM-HIT database

Figure 14 shows the ROC curve of the proposed authentication system when using SVM and KNN classifiers.

From the previous ROC plot, we can observe that the proposed system achieved the highest accuracy when using SVM, which is high performance than KNN classifier. We can also employ the KNN for authentication as it gives an acceptable performance. In our case, we can use automatic hyper-parameter optimization to minimize *ten-*fold cross-validation loss as shown in Figs. 15 and 16. From the Figures and after evaluated 30 functions we observed that the best observed and estimated feasible points are reached when used 19 neighbors and *seuclidean* as a distance function.

Table 5 shows the results of our system using SVM and KNN in terms of Accuracy, Pre, Rec and F-score.

From the previous results, we can observe that the performance of the proposed authentication method using SVM classifier is better than using KNN classifier on the MWM-HIT database. We achieved the best authentication accuracy of 99.4%, Pre of 99.4%, Rec of 100% and F-score of 99.6%. However, the use of KNN classifier is also achieved good authentication accuracy.

To create an even platform for the comparison of ECG authentication systems, the popular online PTB database was used to test the system. Table 6 shows a comparison

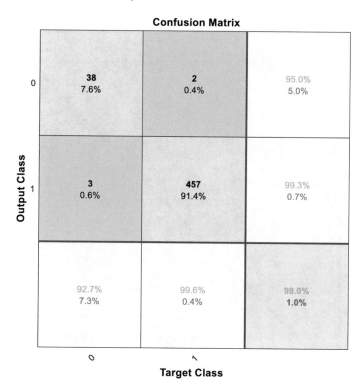

Fig. 13 Confusion Matrix of the proposed system using KNN classifier on MWM-HIT database

of the previous ECG authentication system based on deep learning approaches on the PTB database.

From Table 6, we can observe that the proposed system achieved the highest accuracy comparing to other previous systems. It is worthwhile to mention that the size of the dataset used in the previous study is very limited and obtains low authentication performance comparing with our method. However, the proposed method employed big data and also achieved high performance on limited data with stable authentication accuracy.

6 Conclusion

This research work presented an ECG authentication system based on the deep transfer learning technique. The proposed method achieved a good performance in biometric authentication in different conditions of ECG signals. To evaluate the proposed method, two ECGs databases are employed the MWM-HIT and PTB database. The performance of the system is compared with other ECG authentication methods in Table 6 and the result showed an improved performance. Also,

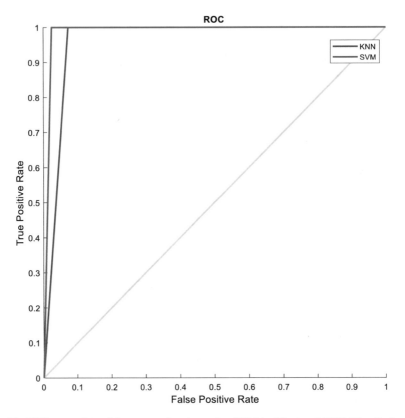

Fig. 14 ROC curve plots of the proposed system using SVM (red line) and KNN (blue line)

Fig. 15 3D plot of the objective function model where *seuclidean* function gives the best estimated values

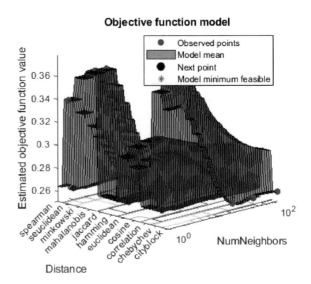

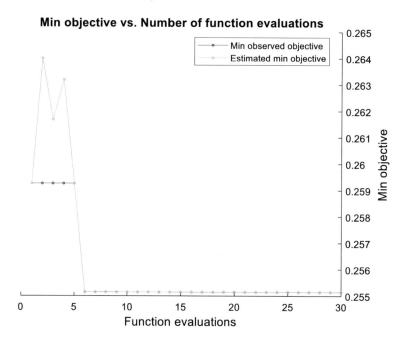

Fig. 16 Number of function evaluations versus the minimum objective

Table 5 The performance of the proposed system using SVM and KNN classifiers

Classifier	Accuracy (%)	Pre (%)	Rec (%)	F-score (%)
SVM	99.4	99.4	100	99.6
KNN	99	99.3	95	97.1

Table 6 Comparison of the performance between the previous authentication methods and the proposed method on PTB database

Authors	Year	Method	Performance
Hammad et al. [41]	2019	Scanning and removing methods CNN	Accuracy: 99.2%
Hammad and Wang [42]	2019	CNN	Accuracy: 96.56% EER: 2.9%
Labati et al. [36]	2019	CNN	EER: 3.5%
Hammad et al. [43]	2018	CNN with Addition fusion	Accuracy: 98.66%
Chu et al. [35]	2019	CNN	EER: 0.59%
Abdeldayem et al. [37]	2019	2-D CNN	Accuracy: 94.9%
Proposed	2021	Pre-train deep models Internal fusion	Accuracy: 99.6%

the time cost of the method is less as compared to previous deep learning methods as we worked on pre-train models and only fine-tuned these models to be suitable for our case. The ECG authentication system produced the best validation accuracy of 99.4% using SVM obtained from the publicly MWM-HIT database. Comparing with other previous ECG authentication systems based on deep learning, the proposed system achieved the highest authentication accuracy. The proposed system is more robust, efficient and can be used for real-time applications. In the future, we can try to use more deep learning algorithms such as Long Short-Term Memory (LSTM) [73–75]. In addition, we can try to authenticate the ECG signals using deep learning on imbalanced data [76].

References

1. Wang, N., Li, Q., Abd El-Latif, A.A., Peng, J., Niu, X.: Multibiometrics fusion for identity authentication: dual iris, visible and thermal face imagery. Int. J. Secur. Appl 7(3) (2013)
2. Wang, N., Li, Q., Abd El-Latif, A.A., Peng, J., Niu, X.: Two-directional two-dimensional modified fisher principal component analysis: an efficient approach for thermal face verification. J. Electron. Imaging 22(2), 023013
3. Wang, N., Li, Q., Abd El-Latif, A.A., Peng, J., Niu, X.: A novel multibiometric template security scheme for the fusion of dual iris, visible and thermal face images. J. Comput. Inf. Syst. 9(19), 1–9
4. Wang, N., Li, Q., Abd El-Latif, A.A., Peng, J., Niu, X.: An enhanced thermal face recognition method based on multiscale complex fusion for Gabor coefficients. Multimed. Tools Appl. 72(3), 2339–2358 (2014)
5. Wang, N., Li, Q., Abd El-Latif, A.A., Yan, X., Niu, X.: A novel hybrid multibiometrics based on the fusion of dual iris, visible and thermal face images. In: 2013 International Symposium on Biometrics and Security Technologies, pp. 217–223. IEEE (2013)
6. Peng, J., Li, Q., Abd El-Latif, A.A., Niu, X.: Finger multibiometric cryptosystems: fusion strategy and template security. J. Electron. Imaging 23(2), 023001 (2014)
7. Peng, J., Li, Q., Abd El-Latif, A.A., Niu, X.: Linear discriminant multi-set canonical correlations analysis (LDMCCA): an efficient approach for feature fusion of finger biometrics. Multimed. Tools Appl. 74(13), 4469–4486 (2015)
8. Hammad, M., Wang, K.: Fingerprint classification based on a Q-Gaussian multiclass support vector machine. In: Proceedings of the 2017 International Conference on Biometrics Engineering and Application, pp. 39–44 (2017)
9. Kang, B.J., Park, K.R.: Multimodal biometric method based on vein and geometry of a single finger. IET Comput. Vis. 4(3), 209–217 (2010)
10. Gad, R., Abd El-Latif, A.A., Elseuofi, S., Ibrahim, H.M., Elmezain, M., Said, W.: IoT security based on iris verification using multi-algorithm feature level fusion scheme. In: 2019 2nd International Conference on Computer Applications & Information Security (ICCAIS), pp. 1–6. IEEE (2019)
11. Gad, R., Talha, M., Abd El-Latif, A.A., Zorkany, M., Ayman, E.S., Nawal, E.F., Muhammad, G.: Iris recognition using multi-algorithmic approaches for cognitive internet of things (ciot) framework. Future Gener. Comput. Syst. 89, 178–191 (2018)
12. Rosdi, B.A., Shing, C.W., Suandi, S.A.: Finger vein recognition using local line binary pattern. Sensors 11(12), 11357–11371 (2011)
13. Shaydyuk, N.K., Cleland, T.: Biometric identification via retina scanning with liveness detection using speckle contrast imaging. In: 2016 IEEE International Carnahan Conference on Security Technology (ICCST), pp. 1–5. IEEE (2016)

14. Tuncer, T., Dogan, S.: Novel dynamic center based binary and ternary pattern network using M4 pooling for real world voice recognition. Appl. Acoust. **156**, 176–185 (2019)
15. Tahir, M., Sardaraz, M., Ikram, A.A.: EPMA: efficient pattern matching algorithm for DNA sequences. Expert Syst. Appl. **80**, 162–170 (2017)
16. Hammad, M., Ibrahim, M., Hadhoud, M.: A novel biometric based on ECG signals and images for human authentication. Int. Arab J. Inf. Technol. **13**(6A), 959–964 (2016)
17. Bidgoly, A.J., Bidgoly, H.J., Arezoumand, Z.: A survey on methods and challenges in EEG based authentication. Comput. Secur. 101788 (2020)
18. Griswold-Steiner, I., Matovu, R., Serwadda, A.: Handwriting watcher: a mechanism for smartwatch-driven handwriting authentication. In: 2017 IEEE International Joint Conference on Biometrics (IJCB), pp. 216–224. IEEE (2017)
19. Gafurov, D., Snekkenes, E., Bours, P.: Spoof attacks on gait authentication system. IEEE Trans. Inf. Forensics Secur. **2**(3), 491–502 (2007)
20. Xi, K., Ahmad, T., Han, F., Hu, J.: A fingerprint based bio-cryptographic security protocol designed for client/server authentication in mobile computing environment. Secur. Commun. Netw. **4**(5), 487–499 (2011)
21. Tehranipoor, F., Karimian, N., Wortman, P.A., Chandy, J.A.: Low-cost authentication paradigm for consumer electronics within the internet of wearable fitness tracking applications. In: 2018 IEEE International Conference on Consumer Electronics (ICCE), pp. 1–6. IEEE (2018)
22. Okokpujie, K., Noma-Osaghae, E., Okesola, O., Omoruyi, O., Okereke, C., John, S., Okokpujie, I.P.: Fingerprint biometric authentication based point of sale terminal. In: International Conference on Information Science and Applications, pp. 229–237. Springer, Singapore.
23. Karthikeyan, S., Feng, S., Rao, A., Sadeh, N.: Smartphone fingerprint authentication versus pins: a usability study. Carnegie Mellon University Technical Reports, pp. 14–012 (2014)
24. Altan, G., Kutlu, Y., Yeniad, M.: ECG based human identification using second order difference plots. Comput. Methods Program. Biomed. **170**, 81–93 (2019)
25. Goshvarpour, A., Goshvarpour, A.: Human identification using a new matching pursuit-based feature set of ECG. Comput. Methods Program. Biomed. **172**, 87–94 (2019)
26. Pinto, J.R., Cardoso, J.S., Lourenço, A., Carreiras, C.: Towards a continuous biometric system based on ECG signals acquired on the steering wheel. Sensors **17**(10), 2228 (2017)
27. Alotaiby, T.N., Alshebeili, S.A., Aljafar, L.M., Alsabhan, W.M.: ECG-based subject identification using common spatial pattern and SVM. J. Sens. (2019)
28. El_Rahman, S.A.: Biometric human recognition system based on ECG. Multimed. Tools Appl. **78**(13), 17555–17572 (2019)
29. Barros, A., Resque, P., Almeida, J., Mota, R., Oliveira, H., Rosário, D., Cerqueira, E.: Data improvement model based on ECG biometric for user authentication and identification. Sensors **20**(10), 2920 (2020)
30. Goshvarpour, A., Goshvarpour, A.: Human identification using information theory-based indices of ECG characteristic points. Expert Syst. Appl. **127**, 25–34 (2019)
31. Wang, K., Yang, G., Huang, Y., Yin, Y.: Multi-scale differential feature for ECG biometrics with collective matrix factorization. Pattern Recogn. **102**, 107211 (2020)
32. Liu, J., Yin, L., He, C., Wen, B., Hong, X., Li, Y.: A multiscale autoregressive model-based electrocardiogram identification method. IEEE Access **6**, 18251–18263 (2018)
33. Hammad, M., Luo, G., Wang, K.: Cancelable biometric authentication system based on ECG. Multimed. Tools Appl. **78**(2), 1857–1887 (2019)
34. Zhao, Z., Zhang, Y., Deng, Y., Zhang, X.: ECG authentication system design incorporating a convolutional neural network and generalized S-transformation. Comput. Biol. Med. **102**, 168–179 (2018)
35. Chu, Y., Shen, H., Huang, K.: Ecg authentication method based on parallel multi-scale one-dimensional residual network with center and margin loss. IEEE Access **7**, 51598–51607 (2019)
36. Labati, R.D., Muñoz, E., Piuri, V., Sassi, R., Scotti, F.: Deep-ECG: convolutional neural networks for ECG biometric recognition. Pattern Recogn. Lett. **126**, 78–85 (2019)

37. Abdeldayem, S.S., Bourlai, T.: A novel approach for ECG-based human identification using spectral correlation and deep learning. IEEE Trans. Biom. Behav. Identity Sci. **2**(1), 1–14 (2019)
38. Kim, B.H., Pyun, J.Y.: ECG identification for personal authentication using LSTM-based deep recurrent neural networks. Sensors **20**(11), 3069 (2020)
39. Kim, M.G., Pan, S.B.: Deep learning based on 1-D ensemble networks using ECG for real-time user recognition. IEEE Trans. Industr. Inf. **15**(10), 5656–5663 (2019)
40. Hammad, M., Pławiak, P., Wang, K., Acharya, U.R.: ResNet-attention model for human authentication using ECG signals. Expert Syst. e12547 (2020)
41. Hammad, M., Zhang, S., Wang, K.: A novel two-dimensional ECG feature extraction and classification algorithm based on convolution neural network for human authentication. Futur. Gener. Comput. Syst. **101**, 180–196 (2019)
42. Hammad, M., Wang, K.: Parallel score fusion of ECG and fingerprint for human authentication based on convolution neural network. Comput. Secur. **81**, 107–122 (2019)
43. Hammad, M., Liu, Y., Wang, K.: Multimodal biometric authentication systems using convolution neural network based on different level fusion of ECG and fingerprint. IEEE Access **7**, 26527–26542 (2018)
44. Lugovaya, T.S.: Biometric human identification based on electrocardiogram. [Master's thesis]. In: Faculty of Computing Technologies and Informatics. Electrotechnical University "LETI", Saint-Petersburg, Russian Federation (2005)
45. Goldberger, A., Amaral, L., Glass, L., Hausdorff, J., Ivanov, P.C., Mark, R., et al.: PhysioBank, PhysioToolkit, and PhysioNet: components of a new research resource for complex physiologic signals. Circulation. **101**(23), e215–e220 (2000)
46. Moody, G.B., Mark, R.G.: The impact of the MIT-BIH arrhythmia database. IEEE Eng. Med. Biol. **20**(3), 45–50 (2001) (PMID: 11446209)
47. Bousseljot, R., Kreiseler, D., Schnabel, A.: Nutzung der EKG-Signaldatenbank CARDIODAT der PTB über das Internet. Biomedizinische Technik, Band 40, Ergänzungsband **1**, S317 (1995)
48. Pan, J., Tompkins, W.J.: A real-time QRS detection algorithm. IEEE Trans. Biomed. Eng. **3**, 230–236 (1985)
49. Peris-Lopez, P., González-Manzano, L., Camara, C., de Fuentes, J.M.: Effect of attacker characterization in ECG-based continuous authentication mechanisms for internet of things. Futur. Gener. Comput. Syst. **81**, 67–77 (2018)
50. Zhang, Y., Gravina, R., Lu, H., Villari, M., Fortino, G.: PEA: parallel electrocardiogram-based authentication for smart healthcare systems. J. Netw. Comput. Appl. **117**, 10–16 (2018)
51. Huang, P., Guo, L., Li, M., Fang, Y.: Practical privacy-preserving ECG-based Authentication for IoT-based healthcare. IEEE Internet Things J. **6**(5), 9200–9210 (2019)
52. Zhang, W.Z., Elgendy, I.A., Hammad, M., Iliyasu, A.M., Du, X., Guizani, M., Abd El-Latif, A.A.: Secure and optimized load balancing for multi-tier IoT and edge-cloud computing systems. IEEE Internet Things J. (2020)
53. The Nymi SDK 3.1 (Beta) documentation (2021). https://downloads.nymi.com/sdkDoc/doc-v3.1.5.326-326_5df03a4/index.html#introduction
54. Fushiki, T.: Estimation of prediction error by using K-fold cross-validation. Stat. Comput. **21**(2), 137–146 (2011)
55. Jagtap, S.K., Uplane, M.D.: The impact of digital filtering to ECG analysis: butterworth filter application. In: 2012 International Conference on Communication, Information & Computing Technology (ICCICT), pp. 1–6. IEEE (2012)
56. Rai, H.M., Trivedi, A., Shukla, S., Dubey, V.: ECG arrhythmia classification using daubechies wavelet and radial basis function neural network. In: 2012 Nirma University International Conference on Engineering (NUiCONE), pp. 1–6. IEEE (2012)
57. Sharma, M., Raval, M., Acharya, U.R.: A new approach to identify obstructive sleep apnea using an optimal orthogonal wavelet filter bank with ECG signals. Inform. Med. Unlock. **16**, 100170 (2019)
58. Hammad, M., Maher, A., Wang, K., Jiang, F., Amrani, M.: Detection of abnormal heart conditions based on characteristics of ECG signals. Measurement **125**, 634–644 (2018)

59. Kaur, M., Singh, B.: Comparison of different approaches for removal of baseline wander from ECG signal. In: Proceedings of the International Conference & Workshop on Emerging Trends in Technology, pp. 1290–1294 (2011)
60. Amrani, M., Hammad, M., Jiang, F., Wang, K., Amrani, A.: Very deep feature extraction and fusion for arrhythmias detection. Neural Comput. Appl. **30**(7), 2047–2057 (2018)
61. Simonyan, K., Zisserman, A.: Very deep convolutional networks for large-scale image recognition (2014). arXiv:1409.1556
62. Schmidt-Hieber, J.: Nonparametric regression using deep neural networks with ReLU activation function. Ann. Stat. **48**(4), 1875–1897 (2020)
63. Yuan, Z.W., Zhang, J.: Feature extraction and image retrieval based on AlexNet. In: Eighth International Conference on Digital Image Processing (ICDIP 2016), vol. 10033, p. 100330E. International Society for Optics and Photonics (2016)
64. Li, S., Jiao, J., Han, Y., Weissman, T.: Demystifying resnet (2016). arXiv:1611.01186
65. Haghighat, M., Abdel-Mottaleb, M., Alhalabi, W.: Fully automatic face normalization and single sample face recognition in unconstrained environments. Expert Syst. Appl. **47**, 23–34 (2016)
66. Chang, C.C., Lin, C.J.: LIBSVM: a library for support vector machines. ACM Trans. Intell. Syst. Technol. (TIST) **2**(3), 27 (2011)
67. Alghamdi, A., Hammad, M., Ugail, H., Abdel-Raheem, A., Muhammad, K., Khalifa, H.S., Abd El-Latif, A.A.: Detection of myocardial infarction based on novel deep transfer learning methods for urban healthcare in smart cities. Multimed. Tools Appl. 1–22 (2020)
68. Aburomman, A.A., Reaz, M.B.I.: A novel SVM-kNN-PSO ensemble method for intrusion detection system. Appl. Soft Comput. **38**, 360–372 (2016)
69. Książek, W., Hammad, M., Pławiak, P., Acharya, U.R., Tadeusiewicz, R.: Development of novel ensemble model using stacking learning and evolutionary computation techniques for automated hepatocellular carcinoma detection. Biocybern. Biomed. Eng. **40**(4), 1512–1524 (2020)
70. Dureja, H., Gupta, S., Madan, A.K.: Topological models for prediction of pharmacokinetic parameters of cephalosporins using random forest, decision tree and moving average analysis. Sci. Pharm. **76**(3), 377–394 (2008)
71. Perdana, R.S., Pinandito, A.: Combining likes-retweet analysis and naive bayes classifier within twitter for sentiment analysis. J. Telecommun. Electron. Comput. Eng. (JTEC) **10**(1–8), 41–46 (2018)
72. Tadeusiewicz, R.: Neural networks in mining sciences–general overview and some representative examples. Arch. Min. Sci. **60**(4), 971–984 (2015)
73. Hammad, M., Iliyasu, A.M., Subasi, A., Ho, E.S., Abd El-Latif, A.A.: A Multi-tier deep learning model for arrhythmia detection. IEEE Trans. Instrum. Meas. (2020)
74. Sedik, A., Iliyasu, A.M., El-Rahiem, A., Abdel Samea, M.E., Abdel-Raheem, A., Hammad, M., et al.: Deploying machine and deep learning models for efficient data-augmented detection of COVID-19 infections. Viruses **12**(7), 769 (2020)
75. Sedik, A., Hammad, M., Abd El-Samie, F.E., Gupta, B.B., Abd El-Latif, A.A.: Efficient deep learning approach for augmented detection of Coronavirus disease. Neural Comput. Appl. 1–18 (2021)
76. Hammad, M., Alkinani, M.H., Gupta, B.B., Abd El-Latif, A.A.: Myocardial infarction detection based on deep neural network on imbalanced data. Multimed. Syst. 1–13 (2021)

Overview of Information Hiding Algorithms for Ensuring Security in IoT Based Cyber-Physical Systems

Oleg Evsutin, Anna Melman, and Ahmed A. Abd El-Latif

Abstract Cyber-physical systems are one of the key technological trends in the modern world. However, their use is associated with the need to counter a variety of cyber threats. This review is devoted to information hiding methods and algorithms designed to ensure security in cyber physical systems. This application of embedding methods is fairly new, but it has attracted the attention of many researchers. The main contribution of our review consists in a new classification of methods for embedding information into data transmitted in cyber-physical systems. We show that these methods can be divided into four broad groups. The main feature for this division is the type of data used to embed additional information. Our review shows that the methods of data hiding used in cyber-physical systems have already formed a whole area that markedly differs from classical digital steganography and digital watermarking.

Keywords Information security · Cyber-physical systems · Internet of things · Digital watermarking · Steganography · Data hiding

1 Introduction

Cyber-physical systems have a significant impact on the daily life of a modern person. They are widely used in industry, medicine, ecology, education, etc. Currently, many cyber-physical systems are built using the Internet of Things (IoT) technology. This technology has certain features that make it difficult to use standard solutions used to ensure the security of information systems. First of all, we are talking about the need to

O. Evsutin (✉) · A. Melman
Department of Cyber-Physical Systems Information Security, HSE University, Moscow, Russia

O. Evsutin
Laboratory of Cyber-Physical Systems, V. A. Trapeznikov Institute of Control Sciences of Russian Academy of Sciences, Moscow, Russia

A. A. Abd El-Latif
Mathematics and Computer Science Department, Menoufia University, Shebin El-Koom, Egypt

© The Author(s), under exclusive license to Springer Nature Switzerland AG 2022 81
A. A. Abd El-Latif et al. (eds.), *Security and Privacy Preserving for IoT and 5G Networks*,
Studies in Big Data 95, https://doi.org/10.1007/978-3-030-85428-7_5

save energy consumption in the case of autonomous IoT devices. In particular, this is one of the main reasons for the existence of the so-called "lightweight cryptography" [1–3].

Digital steganography and digital watermarking methods are an alternative to cryptography. These methods allow us to hide additional information in digital objects. Usually, the purpose of using steganographic data hiding is to ensure the confidentiality of information, and the embedding of digital watermarks in digital objects is used in most cases for the purpose of their authentication.

The expediency of using information embedding methods in cyber-physical systems that have energy consumption limitations is explained by the following features of these methods: low computing complexity in general case and the focus on working with redundant data.

This application of embedding methods is fairly new, but it has attracted the attention of many researchers. Therefore, an overview study that systematizes the results obtained to date is useful. Previously, various authors have published review papers on the data hiding in different nature objects [4–8] and methods of IoT data protection [9–13] separately. However, we present an overview of current research in the field of IoT data protection using data hiding techniques for the first time. The main contribution of our review consists in a new classification of methods for embedding information into data transmitted in cyber-physical systems. The proposed classification takes into account not only the type of the cover object, but also the correspondence of the embedding scenario to security problems in the IoT. This allows us to assess the state of the art in the development of steganography and watermarking algorithms for cyber-physical systems and highlight the scenarios that researchers should pay attention to.

The rest of this review is organized as follows. Section 2 describes an overview of information embedding methods. Sections 3, 4, 5 and 6 sequentially describe different groups of methods according to our classification. Section 7 summarizes our study.

2   General Information on Data Embedding Methods

There are two major areas in the field of information hiding in digital objects: digital steganography and digital watermarking [14].

Digital steganography ensures confidentiality of the information that is embedded in digital objects. The main purpose of steganography is to hide the very fact of presence of additional information in a digital object. Transmission of digital objects with embedded information is usually carried out using public communication channels. Often, a secret key or special algorithm parameters known to the sender and receiver of the information are used to increase the security of embedding.

Digital watermarking has different applications. It is used for a variety of authentication tasks. A digital watermark is a mark that allows us, for example, to identify the owner (author) of a digital object, or to confirm the authenticity and integrity of this object. The fact of presence of a digital watermark in a digital object is not always kept secret. However, the main property of digital watermarking is not imperceptibility, but robustness. This property is the basis for the classification of digital watermarking methods. The following types of digital watermarks are distinguished:

- Fragile digital watermarks. They are destroyed by any impact on a digital object. Therefore, their purpose is to control the integrity of the object.
- Semi-fragile watermarks are resistant to a certain set of permitted transforms of a digital object. They are used to protect against falsification of digital objects. For example, compression of a digital image may be a permitted operation because it does not change the content of the digital object. An example of a prohibited operation can be the replacement of individual image fragments with any other fragments.
- Robust digital watermarks are preserved after a variety of transforms applied to the digital object. They are used to confirm the origin of the object. They usually contain information about the owner or creator of a digital object.

Digital objects are understood here as a variety of multimedia objects: digital images, audio data, video data. The transmission medium of these objects was originally computer networks, primarily the global Internet. The emergence of new concepts of information processing and transmission has led to the emergence of new applications for information hiding techniques. Currently, the application of methods for embedding information into digital data in cyber-physical systems is the subject of research by many scientists, which emphasizes the relevance of this direction. The relevant studies can be divided into two large classes:

- Information embedding in multimedia data generated and transmitted in cyber-physical systems.
- Information embedding into data of a different nature, not related to multimedia (sensor data, information and control signals in cyber-physical systems).

Obviously, the first case is applicable not to all cyber-physical systems, but only to those that operate with data of this type. Nevertheless, such systems are not uncommon and such studies are fairly widespread. This is largely due to the fact that the field of information hiding in multimedia has a quite rich history and provides a good basis for new research.

In its turn, the embedding of information into arbitrary data in cyber-physical systems is a broader case, but this direction is represented by a significantly smaller number of studies.

Our research shows that studies in the field of digital steganography and digital watermarking for cyber-physical IoT systems can be divided into four large groups. The following sections describe these groups in ascending order of difference from classical embedding methods.

3 Embedding Information into Media Data in Cyber-Physical Systems Without Defined Use Cases

The first group of studies is devoted to methods of hiding information in digital images (and other digital objects), designed to protect data in cyber-physical systems. This group includes studies that do not have any specific features associated with the declared area of application. The authors propose solutions designed to protect IoT data, but do not indicate any specific application scenarios for their algorithms.

A significant part of such studies is devoted to steganographic information embedding into digital images. They propose a variety of algorithms. But they all have analogues in the field of classical digital steganography. The general scenario of their use is shown in Fig. 1. It does not differ significantly from the classical scenario of steganographic information protection.

In [15], the authors propose three algorithms for steganographic embedding of information into RGB images for the use in the IoT critical infrastructure. These algorithms differ from each other in the number of color channels used for embedding. The positions for embedding the message bits are selected using a secret key.

In [16], the authors give reasons for the use of Least Significant Bits (LSB) based steganography in IoT applications due to its low computing complexity. They propose an approach to improve embedding efficiency using any LSB-based algorithm. This approach is based on a preliminary analysis of the database of cover images. The analysis checks the degree of coincidence between the bits of the embedded message and the bits of the cover image. The secret message should be embedded in the cover with the maximum match.

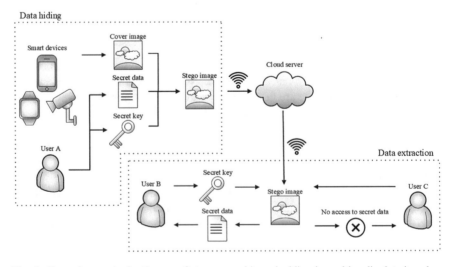

Fig. 1 General scenario for the use of steganographic embedding in multimedia data in cyber-physical systems (for example, images)

In [17], it is proposed to hide the patient information in Magnetic Resonance (MR) images during transmission using the network. The main requirement for the quality of embedding in this case is the imperceptibility of embedding, so that the presence of an embedding does not distort the results of the brain classification on the MR image. The authors claim that classical LSB steganography is well suited for these purposes. Replacement of the bits in the spatial domain does not affect the classification results obtained using a computer-aided diagnosis system.

In [18], the authors propose a steganographic method for protecting Electronic Patient Record (EPR), suitable for transmitting data on the IoT in real time. This method is designed for color images. Parts of the message are embedded in different planes of the cover image using one key. Embedding is performed by replacing two or three LSB. Additionally, a fragile digital watermark, designed to control data integrity, is embedded in the cover image.

In [19], an algorithm for reversible embedding of information into medical images is proposed, which combines the use of interpolation and modular arithmetic. According to the described algorithm, the input image of size $M \times N$ is firstly reduced to the size $M/2 \times N/2$, then the obtained image is enlarged using the pixel repetition method and it becomes a cover image. When embedding information, the container is divided into blocks of 2×2 pixels. The top left pixel of the block does not change during embedding. The difference between the corresponding values of the message elements and the values of the cover pixels modulo 4 or 8 is added to the values of the remaining pixels of the block.

The pre-generation of a cover image is also used in [20]. The original $M \times N$ image is transformed to a $2M \times 2N$ cover image before the embedding using the optimal pixel repetition method. Confidential medical information before the embedding is concatenated with a fragile watermark for authentication at the receiving end. Information embedding is performed using pixel permutation and pixel replacement.

A data hiding scheme based on multidimensional mini-SuDoKu reference matrices is proposed in [21]. The cubic mini-SuDoKu matrix is a three-dimensional fixed-size matrix. This matrix is divided into sub-cubes. Each sub-cube contains eight basic structures of size $2 \times 2 \times 2$. The elements of basic structures are randomly assigned values from 0 to 7. For embedding, a message presented in binary form is split into fragments, which are then hidden in groups of 3 image pixels. Pixel triplets are modified according to the rules depending on the embedded data and cell values of the cubic mini-SuDoKu matrix. The authors also provide an extension of the described approach for the n-dimensional matrix case.

In [22], the authors present a scheme of steganographic information protection in the IoT systems based on vector quantization of digital images. The information embedding procedure is as follows. Firstly, the vector quantization method is applied to the cover image. The size of the blocks is 4×4 pixels. Then, the modules of differences of the pixel block values of the cover image and the image restored after compression are calculated. The message bits are embedded in the pixels of the restored image block by replacing the three LSB. The block index in the codebook is similarly embedded in the first three pixels of the block, the number of bits hidden in a certain block element is embedded in the last pixel.

A scheme for secure storage of multimedia data in the cloud is described in [23]. Before being embedded, the protected information is transformed using the Discrete Rajan Transform (DRT). Then, the transformed multimedia content is embedded in the domain of integer wavelet transform of the cover image and sent in the cloud for storage. Embedding is carried out by changing the difference between pixel pairs using the diamond encoding method. The support vector machine classifier is used during the extraction phase to ensure that the embedded content is restored correctly even after some typical image processing operations.

In [24], a steganographic algorithm based on a Generative Adversarial Network (GAN) is proposed for covert communication in IoT systems. To ensure high efficiency of data hiding, the foreground object region with rich textures is generated on the cover image using the GAN, and then the information is embedded in this specially generated region using the LSB method.

Another GAN-based algorithm is described in [25]. The authors propose a steganography algorithm based on image-to-image translation. CycleGAN is a type of GAN that allows us to transfer image style, for example, turn photos into paintings by famous artists. The authors of the algorithm suggest adding the steganography module and steganalysis module to CycleGAN. The embedding of the secret message occurs in the process of image-to-image translations using the LSB method. Steganalysis is used to judge and supervise the stego image and transferred image.

In [26], the authors propose an image steganography based on evolutionary multi-objective optimization. At the preprocessing stage, the cover image goes through the high-pass filters bank. In the process of information embedding, optimization is applied to maximize the security of the embedding. Embedding is performed in the perturbation locations of the cover image by decreasing or increasing the original pixel by one. Computationally complex operations are implemented on the mobile edge server side, and on the mobile terminal side only information embedding is performed.

In the studies listed above, steganography is an independent method of ensuring information security. However, many papers suggest using steganography in conjunction with cryptography to achieve the best level of security.

For example, in [27], the authors present a model for ensuring the security of textual medical information for transmission in IoT systems. This model combines steganographic information embedding and encryption. At the first stage, confidential data is encrypted using a hybrid encryption scheme based on AES and RSA. At the second stage, the encrypted data is embedded in the discrete wavelet transform (DWT) domain of the cover image by replacing the DWT coefficients.

In [28], a new protocol is proposed for protecting the confidentiality of medical data, which combines the methods of cryptography and steganography. The information is first encrypted using elliptic cryptography, and then it is embedded into the image using the matrix XOR encoding technique. The optimization method called adaptive firefly is used to select the best blocks for embedding data.

The method described in [29] is also based on the combination of steganography and cryptography. On the sender's side, the hash code of a secret message is calculated using the MD5 algorithm. The message itself is divided into four fragments. Each fragment is embedded in one DWT area of the selected digital cover image using the LSB method. The hash code and the obtained stego image are simultaneously transmitted using the communication channel. On the receiver's side, the message is extracted from the stego image, then its hash code is calculated and compared with the transmitted hash code.

In [30, 31], secret sharing schemes based on information hiding theory are proposed.

Some authors propose using audio as a cover object for the covert transmission of information in IoT systems. For example, [32] uses a neural network-based embedding method for this. The embedding model consists of three neural networks: encoder which embeds the secret message in the carrier, decoder which extracts the message, and discriminator which determines the carriers containing secret messages. The authors note that their scheme is lightweight and suitable for smart devices. Another example of audio steganography is presented in [33]. A feature of the proposed algorithm is the choice of optimal positions for embedding secret message bits.

The methods and algorithms presented in our review differ in terms of efficiency. In the current section, we consider the classic data hiding schemes for embedding information in multimedia data, mainly images. The most common quality metrics are Peak Signal-to-Noise Ratio (PSNR), Structural Similarity Index Measure (SSIM), Normalized Cross-Correlation (NCC).

The PSNR metric characterizes the visual similarity of the cover image and the stego image and is calculated by the formula

$$\text{PSNR (dB)} = 10 \times \log_{10}\left(\frac{255^2}{\text{MSE}}\right), \tag{1}$$

$$\text{MSE} = \frac{1}{M \times N} \sum_{i=1}^{N \times M} (C_i - S_i)^2, \tag{2}$$

where $M \times N$ is the height and width of the cover image, C_i is the intensity of the cover image pixel, S_i is the intensity of pixel of stego image.

SSIM is another common metric for assessing the visual similarity of two images. The SSIM value is calculated using the formula

$$\text{SSIM} = \frac{(2\mu_C\mu_S + K_1) \times (2\sigma_{CS} + K_2)}{\left(\mu_C^2 + \mu_S^2 + K_1\right) \times \left(\sigma_C^2 + \sigma_S^2 + K_2\right)}, \tag{3}$$

where μ_C is the mean pixels value of cover image, μ_S is the mean pixels value of stego image, σ_C^2 is the variance of cover image pixel values, σ_S^2 is the variance of stego image pixel values, σ_{SC} is the covariance of both images, K_1 and K_2 are constants.

NCC is used both to assess the similarity of the cover image and the stego image, and to assess the similarity of the extracted secret image with the original secret image. This value is calculated by the formula

$$NCC = \frac{\sum_{i=1}^{M} \sum_{j=1}^{N} C_{ij} S_{ij}}{\sqrt{\sum_{i=1}^{M} \sum_{j=1}^{N} C_{ij}^2} \sqrt{\sum_{i=1}^{M} \sum_{j=1}^{N} S_{ij}^2}} \tag{4}$$

An important performance indicator is the embedding capacity, which measures the amount of additional information hidden in the cover object. The Bit Error Rate (BER) metric shows how many bits of the secret message were extracted incorrectly.

In Table 1, we grouped the studies in the field of steganography for IoT applications. The table shows the main performance indicators and the purpose of use for each method in the IoT environment. Some of the cells in the Performance Indicators column have an N/A value. This means that the authors did not provide any numerical results in their papers or chose such indicators that are not suitable for display in this table. For example, embedding time value is highly dependent on the hardware used, so it would be incorrect to compare it.

As Table 1 shows, many authors describe their solutions as lightweight, which makes them suitable for constrained IoT devices. Many studies are related to the protection of medical data for telemedicine systems.

The following studies are devoted to embedding digital watermarks into data transmitted in cyber-physical systems. On the contrary, they are mostly about frequency embedding, although there are also exceptions. The corresponding scenario (Fig. 2) also has little difference from the classic scenario of embedding digital watermarks.

For example, a scheme for protecting images of DNA microarrays based on the use of fragile watermarking is proposed in [34]. This scheme allows protecting both the whole image and some part of it representing the region of interest. The process of hiding a watermark consists in segmentation, shift, and embedding. Segmentation is used to isolate spots on the image of a DNA microarrays and obtain a raster mask. According to this mask, at the next stage, the pixels are shifted, which consists in decreasing them by one. Then, the elements of the digital watermark are additively embedded into the pixel blocks.

In [35], the authors describe an algorithm for embedding watermarks in images for data exchange in IoT systems. The proposed scheme is based on a random coefficient selection and mean modification approach. Several DCT coefficients randomly selected from two blocks are used to embed information. This ensures that the embedded information is evenly distributed throughout the cover. To create an additional layer of security, the watermark is encrypted before being embedded in the cover image. Embedding of the watermark bits is carried out by changing the ratio between the selected coefficients. The level of change is controlled by a threshold value that affects the robustness of the scheme.

An algorithm for embedding watermarks in medical images based on DWT is proposed in [36]. Before embedding, the cover image undergoes three levels of

Table 1 Media data steganography for cyber-physical systems without defined use cases

Refs. no	Data hiding type	Cover object	Purpose of use	Performance indicators
Bairagi et al. [15]	Steganography	Image	Protecting communication in critical IoT infrastructure	Capacity: 1.41 – 1.97 bpp PSNR: ≈ 51 – 65 dB NCC: 0.99 – 1
Li et al. [16]	Steganography	Image	A lightweight scheme to protect data in IoT systems	N/A
Devi et al. [17]	Steganography	MR image	Patients' personalized data protection for Internet of Medical Things No affecting accuracy of classifying pathological brain	Capacity: 1000–10,000 characters PSNR: 35.67–39.55 dB BER: 0.43–0.92
Parah et al. [18]	Steganography, watermarking	Medical image	Ensuring security for EPR during transit	Capacity: 6–9 bpp PSNR: 37.25–37.68 dB SSIM: 0.7614–0.7803 BER: 0 NCC: 1
Parah et al. [19]	Steganography	Medical/general image	Reversible scheme to protect medical data with low computational complexity	Capacity: 1.5–2.25 bpp PSNR: 39.12–45.49 dB SSIM: 0.9439–0.9868 NCC: 1
Kaw et al. [20]	Steganography, watermarking	Medical image	Reversible scheme to protect EPR for IoT driven e-health	Capacity: 1.25 bpp PSNR: 39.72–43.92 dB SSIM: 0.9119–0.9876 NCC: 1

(continued)

Table 1 (continued)

Refs. no	Data hiding type	Cover object	Purpose of use	Performance indicators
Horng et al. [21]	Steganography	Image	A power saving scheme for security and of multimedia communication among IoT devises	Capacity: 524,288–1,572,864 bits PSNR: 46.36–46.38 dB SSIM: 0.9906–0.9958
Huang et al. [22]	Steganography	Image	Sensitive data leakage protection for IoT networks	Capacity: 404,959–704,569 bits PSNR: 27.55–35.71 dB
Sukumar et al. [23]	Steganography	Image	Content protection for cloud-based storage	Capacity: 65,536–131,072 bits PSNR: 50.11–53.72 dB SSIM: 0.9817–0.9895
Cui et al. [24]	Steganography	Image	Real-time covert communication in IoT	SSIM: 0.9900–0.9980
Meng et al. [25]	Steganography	Image	Covert communication and privacy preserving of the IoT	N/A
Ding et al. [26]	Steganography	Image	Real-time covert communication in IoT	Capacity: 0.1 bpp PSNR: 82.75 SSIM: 1
Elhoseny et al. [27]	Steganography	Medical image	Securing the diagnostic text data transmitting in IoT healthcare systems	Capacity: 15–256 bytes PSNR: 50.52–57.44 dB SSIM: 1 BER: 0 NCC: 1
Khari et al. [28]	Steganography	Image	Protecting medical data during transmission in the IoT	Capacity: 15 bits per cover block PSNR: 70 dB
Yassin et al. [29]	Steganography	Image	Protecting data against attacks during data exchange in IoT environment	PSNR: 81.76–83.01 dB
Yan et al. [30]	Secret sharing	Image	Secure distribution of images	N/A

(continued)

Table 1 (continued)

Refs. no	Data hiding type	Cover object	Purpose of use	Performance indicators
Yan et al. [31]	Secret sharing	Image	Secure distribution of images	N/A
Jiang et al. [32]	Steganography	Audio	A lightweight scheme to protect data in IoT systems	Carrier MSE: 0.0001–0.0096 Secret MSE: 0.0002–0.005 Carrier SNR: 0.9148–6.8470 Secret SNR: −0.2861 to 1.9642
Anguraj et al. [33]	Steganography	Audio	Secure communication for IoT systems	Capacity: 65–85 audio samples PSNR: 47.62–57.17 dB

Watermark embedding

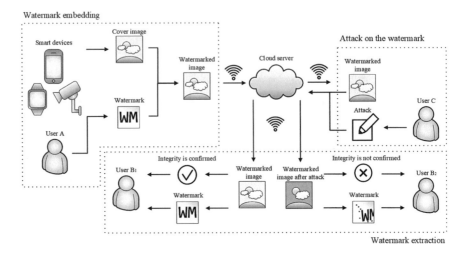

Watermark extraction

Fig. 2 General scenario of application of methods for embedding digital watermarks in multimedia data in cyber-physical systems (for example, images)

Daubechies-3 or Daubechies-9 wavelet transform. The image of the doctor's signature, which is previously transformed in a similar way, is used as a watermark. Data embedding is performed in LL2 sub-band in an additive manner. In [37], the same authors propose a watermarking technique based on a four-level DWT using different wavelet families. These wavelet families are composed of biorthogonal wavelet, reverse biorthogonal wavelet, discrete meyer wavelet, symlet wavelet, and coiflets wavelet transforms.

In [38], a semi-fragile watermarking technique is proposed for color image authentication in power IoT. Cover image preprocessing includes converting the color space of the image from RGB to YUV and performing DCT. A semi-fragile watermark is formed on the base of cover image data for subsequent tamper detection and tamper recovery. Then, a modulo-operation-based watermark embedding is carried out.

A data hiding scheme for image compression with absolute moment block truncation coding (AMBTC) is described in [39]. The authors of this study note that the transmission of compressed images reduces the load on the network. The embedding process consists of three stages. At the intra-block embedding stage, data is embedded into the AMBTC block by modifying the AMBTC parameter set. The swap operator from the direct binary search optimization technology is applied at the second stage to optimize the marked image. At the inter-block embedding stage, additional information is reversibly embedded into a pair of two adjacent blocks.

A zero-watermarking algorithm to protect medical volume data for IoT is proposed in [40]. The Lorenz threedimensional hyperchaotic system is used to encrypt the watermark. As a result, a chaotic watermark is obtained from a meaningful image. After applying 3D double-tree complex wavelet transform and 3D discrete cosine transform (DCT) to original medical volume data, a feature vector is obtained. Embedding of information is performed by applying XOR operation to the feature vector and chaotic watermark. The result of embedding is further used as a secret key. If watermark extraction is required, the XOR operation is applied to the secret key and feature vector obtained from the verified medical data. The Lorenz 3D hyperchaos scrambling algorithm is used to restore the watermark.

In [41], chaotic transformations are used to improve the security of image watermarking schemes.

In general, when transmitting information covertly, video data is often used as covers for embedding. On the IoT, such studies are quite rare. As an example, we can mention the study [42], which proposes a watermarking scheme for secure video transmission in wireless sensor networks (WSN). The authors consider the situation when the video signal in a wireless video surveillance system is substituted by an attacker, and propose a solution to counter this attack. The watermark is embedded in every frame of the MPEG-2 video. Each bit of the watermark is embedded in two DCT coefficients as follows: two auxiliary bits are pre-generated as a result of the action of logical operations AND and OR on the bit of the watermark and one of the bits of the DC coefficient of the DCT block. Then, the auxiliary bits are embedded into the selected DCT coefficients using the LSB method. The least significant bits of all the other DCT coefficients are pseudo-randomly changed to protect against correlation analysis.

In Table 2, we summarize the above watermarking schemes. This research also relates to the field of hiding data in media data without defined use cases. They are characterized by the same performance indicators as steganography methods for embedding information into media data. There are fewer watermark schemes than steganography schemes. Many watermarking techniques are designed to protect medical images in e-healthcare systems.

It is obvious that all these studies are not far from the classical field of information hiding in multimedia data. The proposed methods are potentially applicable in cyber-physical systems, but the scenario of their application is either not specified by the authors, or does not differ from the classical scenarios for the application of embedding methods in computer networks, including the Internet.

4   Embedding Information into Media Data in Cyber-Physical Systems with Defined Use Cases

The next group of studies also covers the classical information embedding in multimedia objects. However, these studies have an important difference from the previous ones. The authors not only declare the applicability of their solutions in cyber-physical systems, but also define the appropriate use cases. Also, many researchers pay attention to the limitations of the proposed methods, which indicates a deeper study in comparison with the studies of the first group.

It should be noted that the proposed application scenarios may significantly differ from each other. Therefore, it is difficult to describe them with some general scheme.

Table 2 Media data watermarking for cyber-physical systems without defined use cases

Refs. no	Data hiding type	Cover object	Purpose of use	Performance indicators
Pizzolante et al. [34]	Watermarking	DNA microarray image	Reversible scheme for protection of consumer genomic data in the internet of living things	Capacity: 128, 256 bits PSNR: 96–101 dB
Hurrah et al. [35]	Watermarking	Image	Preserve privacy and confidentiality of data in an insecure environment of multimedia exchange	Capacity: ≈4500 bits PSNR: 39.93–43.97 dB SSIM: 0.9723–0.9950 BER: 0 NCC: 1
Al-Shayea et al. [36]	Watermarking	Medical image	Medical data protection	NCC: 0.999
Al-Shayea et al. [37]	Watermarking	Medical image	Medical image authentication	PSNR: 121 dB NCC: 1 SSIM: 0.9860–0.9983
Huo et al. [38]	Watermarking	Image	Tampering localization and recovery for image authentication in power IoT	Capacity: 11 bits per 7 quantized DCT coefficients PSNR: ≈41–42 dB
Lin et al. [39]	Watermarking	Image	Image authentication, prevention of forgery attacks, and intellectual property protection Improving the network transmission efficiency	Capacity: ≈80,000–140,000 bits HVS-based PSNR: ≈52–55 dB Mean SSIM: ≈0.9999
Liu et al. [40]	Watermarking	Medical 3D image	Protecting medical volume data for the internet of medical things	NCC: 1

(continued)

Table 2 (continued)

Refs. no	Data hiding type	Cover object	Purpose of use	Performance indicators
Benrhouma et al. [41]	Watermarking	Image	Tamper detection	PSNR: ≈51 dB SSIM: 0.9959–0.9989 True positive rate: 46.77% False positive rate: 0.28%
Wang and Smith [42]	Watermarking	Video	Secure video transportation in WSN	N/A

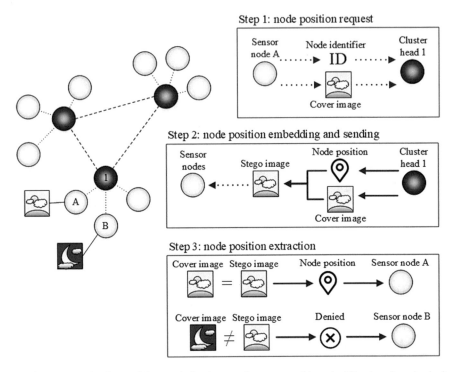

Fig. 3 An example of a special scenario for the use of steganographic embedding in cyber-physical systems

Figure 3 shows an example of such a scenario described in [43]. This paper presents a secure localization scheme for wireless devices based on the use of steganographic information hiding in digital images. For experiments with the proposed scheme, the authors consider a system consisting of multimedia mobile sensor nodes and cluster head nodes. The sensor nodes send position requests to the head node along with their identifiers and cover images. On the cluster head side, position information is embedded into the obtained cover using the LSB method. The stego image is then sent back to sensor nodes, which compare the two images and extract node position information. Only the node containing the original cover image corresponding to the received stego image can receive position information.

Since in the IoT infrastructure the user interacts with various "smart" devices most often via mobile applications, it is important to ensure their security. In [44], a method for protecting Android mobile applications with the help of LSB steganography is proposed. This method is based on the fact that mobile applications contain a large number of PNG images. Steganographic embedding of the main application code in these images allows to protect the code from distortions. According to this scheme, the graphical elements of the interface are displayed in their original form only in case of successful launch. If malicious distortion is introduced into the application, the images are not displayed correctly, which allows the user to detect a fake visually.

The paper [45] describes a two-level data security system. During the phase of information transmission between the sensor node and the authentication server, the message hash code is calculated using the MD5 algorithm and, then, a simple encryption method based on the XOR operation is applied to the data. The resulting hash code and encrypted information are embedded in the image using the LSB method. On the side of the authentication server, information is extracted from the image, the data is decrypted, followed by the calculation of the hash code, which is compared with the hash code extracted from the stego image. If the integrity of the data has been violated, the transmission is blocked. During the phase of data transmission between the authentication server and the cloud, the authors propose to use a more sophisticated MSB-LSB method of steganographic embedding and standard secure encryption algorithms, such as AES.

In [46], the authors present a method for ensuring the security of smart locks. Access control to such a lock is carried out by a smartphone, which communicates with the server using the Bluetooth Low Energy (BLE) protocol. The decision to access is made by the server. The proposed method is aimed at overcoming the vulnerability of this protocol to a man-in-the-middle attack. The smartphone user enters the access key through the app. The access key is encrypted using the AES algorithm, and then the encrypted key is embedded in the image. The image is sent to the server using the BLE protocol. On the server side, an encrypted access key is extracted and decrypted. Then, it is checked whether the received access key matches the correct one, and a decision is made whether to open the smart lock or not.

A similar scenario is proposed in [47]. A snapshot of a person's face at the entrance to the house is used as a container for additional information. Before sending to the server, a watermark is embedded in the image, and the front door opens only if the watermark in the resulting image is correct. To improve the security of embedding, the authors use pre-processing of the cover image with genetic algorithm coding.

In [48], it is proposed to use steganography to improve the security of biometric authentication for IoT systems. The secret key used in the cancelable iris-based authentication system is proposed to be hidden in images separately transmitted to the server in order to prevent intruders from accessing it.

In [49], an algorithm for embedding digital watermarks into digital images in wireless multimedia sensor networks based on DWT is proposed. The authors point out the problem of destruction of digital watermarks due to the noise in wireless communication channels and packet loss. They claim their algorithm to be a solution to this problem. The main idea is to link the embedding parameters to the characteristics of the network, in particular to the packet loss rate. Adaptive parameters can control the redundancy of the embedding. In addition, the authors of [49] solve the problem of choosing the data transmission channel parameters that provide the least energy consumption.

In [50], the authors propose a scheme for secure transmission of images in telemedicine systems. Encrypted confidential images are embedded in images whose content is not confidential. Additionally, a fingerprint (perceptual hash) of a confidential image is embedded in the cover image for the purpose of its subsequent authentication. A distinctive feature of this scheme is tracking the order of image transmission. To do this, the authors introduce the concept of an image fingerprint by analogy with the concept of block chains that underlie the blockchain technology.

In [51], the authors present a digital medical image service management model designed to operate with medical images obtained using VR and AR. Steganography is used to hide confidential patient information that is previously encrypted using RSA and AES algorithms. Information embedding is implemented by concatenation of multimedia data and an encrypted message. A special feature of the study is the description of a proxy signature-based protocol that regulates the exchange of keys and the transmission of confidential data.

In [52], a combination of data aggregation and data hiding is used for efficient transmission in a network of aerial video synthetic aperture radar (ViSAR) vehicles. The authors propose an approach to interpolation-based data hiding using DCT, which can significantly reduce the file size for wireless transmission. After applying DCT to the host frame, DCT coefficients scale into an interval of [0, 255], then, they are quantized, resulting in a new host frame with quasi-sparse spatial distribution. Then, its interpolated version is created, an error image is formed, and using the histogram of the error image, the key embedding parameters are calculated. The data is embedded in the quasi-sparse host frame according to the calculated parameters.

In [53], the authors demonstrate the implementation of a data hiding algorithm for 32-bit reduced instruction set computer microcontrollers for IoT platforms. The embedding cover and the secret message are small halftone images. The pixels of the cover image and the secret image are stored as a byte stream in sequential address cells of the built-in flash memory and SRAM of the microcontroller, respectively. The embedding is carried out using the LSB method. The stego image pixels are stored as a byte array in the SRAM of the microcontroller and serially transmitted asynchronously via on-chip Universal Synchronous/Asynchronous Receiver/Transmitter (USART) of the microcontroller.

In [54], a novel secret sharing scheme for image is proposed based on Shamir's polynomials with steganography the IoT system, where IoT devices could share images seamlessly between them over cloud. In this proposed method, secret shares are generated from the cover image and secret information, and the shares of IoT services are embedded back to the cover image based on on 24-ary notational system, which creates stego-shadow images.

The study [55] is somewhat specific, as it deals with the embedding of hidden attachments in images used in printed products. However, the authors quite clearly define the applications of the proposed approach in IoT systems and the corresponding application scenarios, in particular, to ensure data authentication in order to protect products from forgery. Therefore, the study corresponds to the topic of this review. It should also be noted here that the authors talk about steganographic embedding, however, they focus on the robustness, which is typical for embedding digital watermarks.

One more research direction should be noted that goes beyond the classical field of data hiding. This direction is associated with embedding data into media objects using the quantum computing paradigm. Relevant studies have a forward-looking nature. Advances in quantum technologies are expected to usher in the quantum era whence there will be adequate computing power to tamper with the best of traditional algorithms. Some results in the field of quantum embedding algorithms with proposed use cases in cyber-physical systems can be found in [56–59].

In Table 3, we show basic information about the studies described above. As follows from the table, most of the methods described in this section are related to the field of steganography. The cover objects for additional information are digital images, but the options for the purpose of use are more diverse than in Sect. 3. The most common use for this type of methods is authentication in IoT systems.

Table 3 Embedding information into media data in cyber-physical systems with defined use cases

Refs. no	Data hiding type	Cover object	Purpose of use	Performance indicators
Tondwalkar and Vinayakray-Jani [43]	Steganography	Image	Secure localization for wireless devices	N/A
Kim et al. [44]	Steganography	Image	Protecting the code of Android mobile applications from distortions	Capacity: 2756–4944 bytes
Das et al. [45]	Steganography	Image	Secure data transfer in smart IoT environment	Capacity: 1–99 characters
Bapat et al. [46]	Steganography	Image	Overcoming the vulnerability of BLE protocol to a man-in-the-middle attack for ensuring the security of smart locks	N/A
Zaichenko and Sineva [47]	Watermarking	Image	Secure user's face image transfer to authentication server	Capacity: 41,616 bits PSNR: 51.61–51.89 dB
Yang et al. [48]	Steganography	Image	Security improvement of iris-based biometric authentication for IoT systems	Equal error rate: 1.66–4.78%
Wang [49]	Watermarking	Image	Multimedia authentication in wireless multimedia sensor networks	N/A
Peng et al. [50]	Steganography	Image	Secure transmission of images in telemedicine systems	PSNR of reconstructed image: ≈33 dB

(continued)

Table 3 (continued)

Refs. no	Data hiding type	Cover object	Purpose of use	Performance indicators
Yoon-Su and Seung-Soo [51]	Steganography	Image	Integrity of user multimedia image information processed through special medical equipment using VR	Efficiency in healthcare information management: $\approx$0.7–0.9
Khosravi and Samadi [52]	Steganography	ViSAR frame image	Data aggregation for IoT-enabled ViSAR sensor networks	Capacity: 2950–29,800 bits PSNR: 48.68–51.41 dB SSIM: 0.9935–0.999
Janakiraman et al. [53]	Steganography	Image	Lightweight security scheme for 32-bit reduced Instruction set computer microcontrollers used in IoT platforms	Capacity: 1.49–1.52 bpp PSNR: $\approx$46 dB SSIM: 0.9998 NCC: 0.999
Li et al. [54]	Secret sharing, steganography	Image	Secure distribution of images	Capacity: 65,536 bits PSNR: $\approx$46.5–47.0 dB
Pu et al. [55]	Steganography	Images used in printed products	Data authentication to protect products from forgery	Information entropy: 0 Contrast: 0.6537–0.8717
bd-El-Atty et al. [56]	Steganography	Medical image	Secure transmission of images on cloud-based e-healthcare platforms	Capacity: 2/8 bpp PSNR: 44.01–44.43 dB SSIM: 0.9318–0.9689 NCC: 0.9991–0.9999
El-Latif et al. [57]	Steganography	Medical image	Secret images transfer in cloud system	Capacity: 2/8 bpp PSNR: 42.5–54.0 dB SSIM: 0.988–1 NCC: 0.995–1
Peng et al. [58]	Steganography	Image	Secure distribution of images	Capacity: 2 bpp PSNR: 48.60 SSIM: 0.9964 NCC: 0.996
EL-Latif et al. [59]	Steganography	Image	Secure distribution of images	Capacity: 2/8 bpp PSNR: 41.51–44.49 dB SSIM: 0.9053–0.9756

Thus, the studies belonging to the second group can be classified in two ways. On the one hand, they actually represent methods that work with media data. This allows them to be classified as a classical field of data hiding. But, on the other hand, the authors of the mentioned studies pay attention to the development of scenarios for the application of their methods in cyber-physical systems. And this is a significant difference from the classical field of data hiding.

5   Embedding of Information into Sensor Data in Cyber-Physical Systems

The next group of studies is devoted to the embedding of information in data generated and transmitted in cyber-physical systems and not related to multimedia. This is data generated by a variety of sensors that are present in cyber-physical systems. Quite a lot of studies are devoted to the embedding of information into the data of WSN. A generalized scheme for applying embedding methods to sensor data in cyber-physical systems is shown in Fig. 4. Not all of the studies belonging to the third group fully fit this scheme. Nevertheless, most of them correspond to it.

It should be noted that most of the studies in this group are devoted to digital watermarking. Research in the field of steganography is also presented, but in smaller numbers. This can partly be explained by the fact that non-media data does not provide such embedding capacity as digital images, and even more as video. At the same time, the problem of data authentication, which is solved using digital watermarking, is relevant for data of any type.

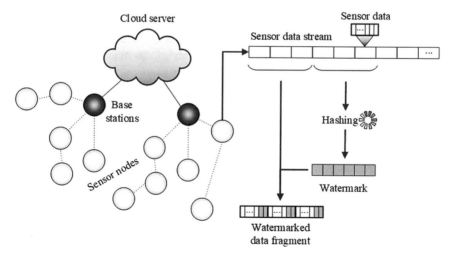

Fig. 4 Scheme of embedding digital watermarks into the data of WSN

The paper [60] is devoted to steganographic information hiding in non-multimedia data in IoT systems. It presents two methods for concealing a driver misbehavior report transmitted between intelligent vehicle sensors on vehicular ad hoc networks (VANETs). VANETs are designed to improve road safety, and their sensors provide real-time information on the status of road traffic. One of the methods suggested in the article is based on LSB steganography. A beacon signal, which is a message that contains the current state of the sender in terms of position, speed, etc., is used as a cover to hide information. Since the accuracy of sensors is limited, there is noise in the transmitted data that can be used for effective steganographic hiding of information.

Another similar example is presented in [61]. It devoted to the aggregation and protection of data confidentiality in body sensor networks designed for continuous monitoring of health status. There are two types of sensors in such networks. Some of them collect data that is not confidential, for example, body temperature and movement sensors. Other sensors collect confidential data, such as an ECG sensor. According to the proposed scheme, such a sensor carries out lossless compression of confidential data. Then it receives data from "ordinary" sensors. To reduce the load on the network, packet combination technology is applied to this data, and several packets are combined into one. Compressed confidential information is embedded into it using the XOR operation. The sequence of packets with "ordinary" data within the combined packet, which will serve as covers for embedding, is determined based on various parameters such as the least significant bits of packets, the number of bits in the first compressed packet, and others using a hash function.

In [62], the authors propose a transmission framework for electrocardiogram (ECG) data. A convolutional neural network based ECG classification algorithm is designed to improve security and reduce network load. An unequal steganography embedding algorithm is proposed for the secure transmission of information in DWT coefficients of physiological signals from sensor nodes to a server. Embedding is carried out in detail coefficients of the wavelet-domain in an additive-multiplicative manner. The embedding strength depends on the importance of the confidential data. The most important data has a higher embedding strength and is transmitted first, while the least important secret data with the least embedding strength is transmitted last.

Another example of steganographic embedding of information in non-media data in the IoT infrastructure is presented in [63]. Its authors propose to hide messages in the MAC layers of ZigBee protocol. A secret message can be hidden in the reserved field bits of data frames. For example, the frame control field can hide 3 and 2 bits of secret message in the 7–9th and 12–13th bits respectively. Information encryption is used to increase the level of security.

A significant part of the studies related to this group deals with the embedding of digital watermarks into the WSN data.

In [64], watermarking is used for data authentication. This study proposes an original approach to the aggregation of sensor data. According to this approach the set of values coming from each group of sensors is represented in the form of a matrix.

The authors of the study propose to consider each of these matrices as a pseudo-image. Since nearby sensors generally fix close values, such a pseudo-image will have spatial redundancy, like an ordinary image. The aggregation consists in JPEG-like compression of the generated pseudo-image. Before compression, a watermark is embedded in the pseudo-image using the direct spread spectrum sequence, in the course of which one bit of the watermark is embedded in a block of adjacent "pixels". The embedding operation consists in changing the values of the "pixels" in the block to small values under the control of a pseudo-random sequence.

In [65], the authors propose a method for checking the integrity of data from various sources, for example, health or environmental monitoring sensors. This method is based on embedding watermarks in data streams. Like the study mentioned above, this study also deals with pseudo-images. To embed a watermark, a spread spectrum method using orthogonal codes is applied. Due to the proposed method, the integrity of the data streams can be verified by extracting the watermark, even if the data goes through several stages of the aggregation process. The proposed water-marking scheme preserves the natural correlations that may exist between several data streams.

In [66], the same authors propose a technology for obfuscating sensor data based on the use of watermarking. A spread spectrum technique is also applied to embed a watermark. Obfuscation is implemented by adding a scaled watermark to the data. Unlike traditional watermarking schemes, where a watermark should be invisible, the amplitude of the watermark for obfuscation purposes should be as large as possible.

An online watermarking scheme to protect the rights to trajectory streams is proposed in [67]. The basic idea is to embed the watermark in a sequence of values that define the distances between pairs of object locations. To define the spatial boundaries of the data stream, the authors use a processing window. As more input data arrives, older data is replaced by newer data. To embed a watermark, it is necessary to identify the locations of objects in the data stream by moving the processing window. If two consecutive object locations are displayed in the same processing window, the distance between the locations is calculated. The watermark bit is then embedded in this distance by changing the bits in a predetermined position.

In [68], another method for embedding watermarks in positioning LiDAR data is described. The authors propose a watermarking scheme, which can be used for copyright protection and data source tracking. To embed a watermark, a vector of marker positions is first determined, i.e. positions where the watermark bits will be embedded using a pseudo-random generator. The embedding is carried out in a circular area around the marker positions, which is split into smaller areas evenly distributed in the circle. The DCT is applied to the distance vector, which is calculated on the base of points in the obtained areas. The embedding of the watermark is done by modulating the last DCT coefficient.

In [69], the authors present a method that allows to link datasets with their source of origin. To do this, they propose to embed additional information about its origin (belonging) into the dataset. The method is designed to work with arbitrary sets of the same data type with redundancy. The origin label is a fixed-length binary sequence. This sequence is split into partially overlapping parts, which are written into n least significant bits of the data set elements. Two additional bits are used for checking: the parameter check bit contains the convolution of the most significant bits of the data element with the value of the number of the origin label parts, the label check bit contains the convolution of the most significant bits of the data element with the value of the label itself.

In [70], an algorithm for embedding digital watermarks into the data of WSN is presented. The main purpose of this algorithm is to provide protection against an attack aimed at cloning sensor nodes. Embedding is based on a transform similar to simple XOR cipher. The algorithm is lightweight, which claims to be its advantage.

These algorithms embed digital watermark elements into sensor values in a sequential and independent manner, since it does not depend on these sensor values or some of their characteristics. However, the idea of generating a digital watermark, depending on the protected data itself, is quite common both for classical methods and algorithms of digital watermarking, and for the considered problem area of data protection in WSN and IoT systems.

In a simpler case, digital watermark elements are generated only based on the values of the sensor data elements. An example is the embedding scheme presented in [71]. According to this scheme, a digital watermark bit embedded in the next sensor value is generated based on several previous sensor values.

In [72], the authors propose a scheme for ensuring the safety of data received from sensors, based on the combined use of watermarking and the compressed sensing (CS) method, which is designed to restore a complete signal from its sparse or compressed representation. On the sensor side, the watermark is generated from the protected data using a hash function. Then, the watermark is embedded in the original signal, whose elements are added with empty symbols depending on the value of the watermark bit. The wireless channel transmits CS-sparse data to the base station. The base station decoder restores the complete signal, and the watermark is extracted from the cover. To check the data integrity, the extracted digital watermark is compared with the digital watermark obtained on the basis of the received data.

In [73], a data hiding technique using the CS method is also used. On the sender side, a low-dimension watermark is embedded in measurements. On the receiver side, a CS-based watermark decryption/reconstruction engine is used. In this case, the watermark is used as an electronic signature to verify legitimate users and protect against DoS attacks. To ensure the correct operation of the described scheme in real time, the multiple-indices updating algorithm and the corresponding very large scale integration architecture are used.

In [74], the authors propose a method for ensuring the integrity of sensor data by grouping data received from sensors to create and embed watermarks. Groups of a variable number of data elements are formed by means of their concatenation. The MD5 hash function is used to generate a watermark. In order to calculate this function, it is required to combine two adjacent groups. The watermark is embedded in the first of two sequential groups using the LSB method.

In [75], a similar grouping of data flow elements is applied. The authors took as a basis the idea of watermarking algorithms for digital images using pixel prediction errors. The sensor node groups the flow data, and two adjacent non-overlapping groups form an authentication group. Some of the sequential data elements are used to generate a watermark. An MD5 hash function is applied to each of these elements and all results are combined by means of XOR operation. The rest of the authentication group data elements are used to embed the watermark. The watermark bit is hidden in each of these elements by addition with the corresponding doubled element, reduced by the amount of prediction error. The prediction error is corrected after changing each element.

This approach has certain advantages as compared to the independent embedding of digital watermark elements, but it introduces a synchronization problem. If the order of the data elements is broken when the message is received, it will lead to errors when extracting the digital watermark even if there is no active intruder on the communication channel. Some way of solving this problem is presented in [76]. The authors propose a two-chain digital watermarking scheme. According to this scheme, the sensor data is divided into groups of variable length, depending on the key. The generation and embedding of digital watermark chains is carried out for pairs of adjacent groups. One chain of digital watermarks serves to authenticate the sensor data itself. The second chain of digital watermarks encodes the separators between groups and ensures synchronization between the sender and receiver of the data.

In a more complex case, not only the values of sensor quantities, but also some of their characteristics are involved in the generation of a digital watermark. In [77], the research is devoted to the problem of authenticating data coming from IoT devices. For this purpose, it is proposed to extract stochastic characteristics of data streams and generate digital watermarks on their basis. A spread spectrum technique is used as a method for embedding digital watermarks in the data stream. In [78], the authors also deal with the use of various characteristics of the captured data in the generation of a digital watermark: data length, frequency of occurrence and capture time. In [79], a digital watermark is generated based on information about collisions of the CSMA / CA protocol and serves to repel an attack of sensor node cloning. In addition, a distinctive feature of this study is the way of presenting sensor data. They form a matrix similar to a digital image. In the general case, such a solution makes it possible to apply, when working with sensory data, approaches that have successfully proven themselves in relation to digital images.

Table 4 summarizes the work discussed in this section. It should be noted that the performance metrics vary significantly. It depends on the main goal of a particular study, as well as on the attack model considered by the authors. In the case of watermarking, detection probability is often used as a performance indicator. In Table 4, this value is presented in fairly wide ranges. This is due to the varying conditions of the experiments conducted by the authors of the studies. We do not give the best conditions in each individual case, since the experiments themselves can be organized in significantly different ways, as a result of which their unified description turns out to be difficult.

This section shows the evolution of methods for embedding information in sensor data in the course of time. The simplest solutions in this field are to embed message or watermark bits into separate sensor values using simple operations like LSB. More

Table 4 Embedding of information into sensor data in cyber-physical systems

Refs. no	Data hiding type	Cover object	Purpose of use	Performance indicators
de Fuentes et al. [60]	Steganography	Beacon message of SAE J2735 standard	Covert reporting of misbehaving vehicles	Capacity: 13 bits/beacon
Ren et al. [61]	Steganography	Combined packages of heterogeneous sensor data	Protecting data privacy for body sensor networks	N/A
Sahu et al. [62]	Steganography	Sensor data	Secure transmission of patient information for remote health monitoring	Cover-stego correlation: 95% Capacity: 0.2–4.0 bpp
Hussain et al. [63]	Steganography	ZigBee protocol frames	Enhancing security in ZigBee protocol	N/A
Zhang et al. [64]	Watermarking	Sensor data	Authentication of sensor data	Detection probability: 0.3–1.0
Panah et al. [65]	Watermarking	Sensor data	Authentication of sensor data	Detection probability: 0.6–1.0
Yavari et al. [66]	Watermarking	IoT data	Obfuscation of IoT data to ensure its privacy (using the example of medical data)	N/A
Yue et al. [67]	Watermarking	Trajectory streams data	Rights protection of trajectory streams	BER: 0.45–0.0
Lipuš and Žalik [68]	Watermarking	LiDAR data	Copyright protection and the data source tracking	Detection probability: 0.3 – 1.0
Chong et al. [69]	Watermarking	Time-series datasets	Time-series data source authentication	Detection probability: 0.24–1.0
Hoang et al. [70]	Watermarking	Sensor data	Sensor data authentication in WSN WSN node cloning protection	N/A
Xiao and Gao [71]	Watermarking	Sensor data	Authentication of individual elements of sensor data	Total false positive and false negative rate: 2.34–4.47

(continued)

Table 4 (continued)

Refs. no	Data hiding type	Cover object	Purpose of use	Performance indicators
Wang et al. [72]	Watermarking	Sensor data	Authentication of sensor data	N/A
Chen et al. [73]	Watermarking	Sensor data	Sensor data authentication in WSN (using electrocardiography signals)	PSNR: 22 dB
Kamel and Juma [74]	Watermarking	Sensor data	Sensor data authentication in WSN	N/A
Shi and Xiao [75]	Watermarking	Sensor data	Sensor data authentication in WSN using reversible watermarking	False negative rate: 0.09–0.01 False positive rate: 0.07–0.01
Wang et al. [76]	Watermarking	Sensor data	Sensor data authentication in WSN using reversible watermarking	False positive rate: 0.06–0.01
Ferdowsi and Saad [77]	Watermarking	IoT data	IoT data authentication	BER: 10^{-8}
Hameed et al. [78]	Watermarking	Sensor data	Sensor data authentication in WSN	N/A
Nguyen et al. [79]	Watermarking	Sensor data	Sensor data authentication in WSN WSN node cloning protection	Detection probability: 0.0–1.0

complex solutions take into account the peculiarities of information transmission in cyber-physical systems. Their authors focus on ensuring that information hiding methods help to improve the security of cyber-physical systems, not only in theory, but also in practice.

6 Embedding Information into Analog Signals in Cyber-Physical Systems

The algorithms presented in the previous section work with digital watermarks, which are binary sequences. In addition, there is a class of studies devoted to the embedding of watermarks in analog signals (in particular, in modulated signals) for solving problems of signal or source authentication. The solutions that can be

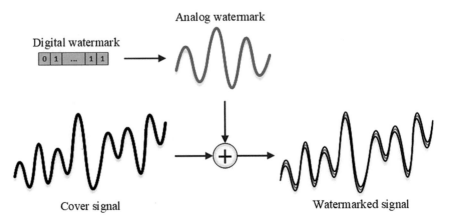

Fig. 5 General scheme of embedding digital watermarks into analog signals in control systems

found in these studies are conceptually similar to the solutions for embedding digital watermarks. The difference lies only in the form of the signal presentation and, as a consequence, in the methods of its processing. An example of the corresponding scheme is shown in Fig. 5.

Several studies are devoted to radio frequency (RF) watermarking. In these studies, the watermarks are embedded in the radio frequency signal. Some RF watermarking techniques work with the watermark as a binary sequence, some involve additional transformations. In the first case, the form of the cover signal changes depending on the bits of the watermark. Several similar schemes are investigated in [80] and compared to each other in terms of channel capacity. In the second case, the watermark is transformed into an analog signal. This signal in some way overlaps the cover signal, resulting in the formation of a watermarked signal. Study [81] is an example of scheme for secure sender authentication in NB-IoT systems. The main advantage of the proposed scheme is the increased reliability due to the elimination of mutual interference between the useful signal and the watermark signal.

In certain cases, watermarking serves to repel specific types of attacks. The studies [82, 83] are devoted to the identification of replay attacks aimed at control systems. This means an attempt by an intruder to interfere with the control of the system by replaying previously captured data sequences. In [83], this problem is considered in relation to networked control industrial systems. The main contribution of this study is not the embedding algorithm, but the strategy for using this algorithm to protect against an intruder.

In [84], the authors present an algorithm for embedding reversible watermarks into signals transmitted in industrial control systems with hard real-time. The authors point to the shipboard control system as a priority area of application. Embedding is additive and is carried out under the control of a secret key, which must first be transmitted using a secure communication channel. The proposed algorithm allows detecting attacks aimed at signal delay and distortion.

Table 5 Embedding information into analog signals in cyber-physical systems

Refs. no	Data hiding type	Cover object	Purpose of use	Performance indicators
Xu and Yuan [80]	Watermarking	Modulated radio frequency signal	Physical layer authentication for radio frequency networks	BER: 10^{-4} to 10^{-1} Capacity: 0.5–1.5 bit/symbol
Huang and Zhang [81]	Watermarking	Modulated radio frequency signal	Physical layer authentication for radio frequency networks More reliable and secure services for practical NB-IoT systems	BER: 10^{-5} to 10^{-1} Capacity: 0.2–2 bit/symbol Spectral efficiency: 1.68 bit/s/Hz
Mo et al. [82]	Watermarking	Sensor signal	Authentication the correct operation of a control system	Detection probability: 0.35–0.80
Rubio-Hernan et al. [83]	Watermarking	Sensor signal	Protection from integrity attacks against cyber-physical systems	Detection ratio: 0.65
Song et al. [84]	Watermarking	Sensor signal	Hard real-time data integrity validation on resource-limited embedded systems	Time of embedding/validation on one sample data point: 2.8 μs

Table 5 summarizes the studies reviewed. In the case of embedding watermarks into a modulated radio frequency signal, the main performance metric is BER. This metric shows the noise immunity of a watermarked signal. This differs from the classical case, when we estimate the watermark resistance to various attacks using this value. In the case of embedding watermarks into sensor signals in automatic control systems, the authors use metrics indicating the probability of successful watermark detection. In automatic control systems with hard real time, the speed of embedding and extracting a watermark comes first.

As can be seen, there are not many similar studies presented in this section. However, their number is gradually increasing. This allows us to talk about the emergence of a new promising area of research. This direction differs both from the classical field of information hiding in multimedia data, and from the growing field of information hiding in digital sensor data in cyber-physical IoT systems.

7 Conclusion

This review focuses on information hiding methods designed to ensure security in cyber-physical systems. This is a fairly new direction in the field of data hiding. It has become more relevant in recent years due to the ubiquitous spread of IoT and related technologies. The main contribution of our review consists in a new classification of methods for embedding information into data transmitted in cyber-physical systems. We point out that these methods can be divided into four broad groups. The main feature for this division is the type of data used to embed additional information. Our review shows that the methods of data hiding used in cyber-physical systems have already formed a whole area that markedly differs from classical digital steganography and digital watermarking. In addition, there is a clear tendency towards the formation of new directions of research. In particular, methods of embedding information into analog signals in control systems began to develop. We can say that the spreading of cyber-physical systems had a positive impact on the field of information embedding and stimulated a new turn of development in this field of science.

Acknowledgements This work was funded by Russian Science Foundation, grant number 19-71-00106.

References

1. Singh, S., Sharma, P.K., Moon, S.Y., Park, J.H.: Advanced lightweight encryption algorithms for IoT devices: survey, challenges and solutions. J. Ambient Intell. Hum. Comput. 1–18 (2017). https://doi.org/10.1007/s12652-017-0494-4
2. Lara-Nino, C.A., Diaz-Perez, A., Morales-Sandoval, M.: Elliptic curve lightweight cryptography: a survey. IEEE Access. **6**, 72514–72550 (2018). https://doi.org/10.1109/ACCESS.2018.2881444
3. Dhanda, S.S., Singh, B., Jindal, P.: Lightweight cryptography: a solution to secure IoT. Wirel. Pers. Commun. **112**, 1947–1980 (2020). https://doi.org/10.1007/s11277-020-07134-3
4. Sahu, A.K., Sahu, M.: Digital image steganography and steganalysis: a journey of the past three decades. Open Comput. Sci. **10**, 296–342 (2020). https://doi.org/10.1515/comp-2020-0136
5. Hua, G., Huang, J., Shi, Y.Q., Goh, J., Thing, V.L.: Twenty years of digital audio watermarking—a comprehensive review. Signal Process. **128**, 222–242 (2016). https://doi.org/10.1016/j.sigpro.2016.04.005
6. Asikuzzaman, M., Pickering, M.R.: An overview of digital video watermarking. IEEE Trans. Circuits Syst. Video Technol. **28**, 2131–2153 (2017). https://doi.org/10.1109/TCSVT.2017.2712162
7. Tian, J., Xiong, G., Li, Z., Gou, G.: A survey of key technologies for constructing network covert channel. Secur. Commun. Netw. (2020). https://doi.org/10.1155/2020/8892896
8. Panah, A.S., Van Schyndel, R., Sellis, T., Bertino, E.: On the properties of non-media digital watermarking: a review of state of the art techniques. IEEE Access **4**, 2670–2704 (2016). https://doi.org/10.1109/ACCESS.2016.2570812
9. Lee, I.: Internet of things (IoT) cybersecurity: literature review and IoT cyber risk management. Future Internet **12**, 157 (2020). https://doi.org/10.3390/fi12090157

10. Waheed, N., He, X., Ikram, M., Usman, M., Hashmi, S.S., Usman, M.: Security and privacy in IoT using machine learning and blockchain: threats and countermeasures. ACM Comput. Surv. (CSUR). **53**, 1–37 (2020). https://doi.org/10.1145/3417987
11. Al-Naji, F.H., Zagrouba, R.: A survey on continuous authentication methods in internet of things environment. Comput. Commun. (2020). https://doi.org/10.1016/j.comcom.2020.09.006
12. Alwarafy, A., Al-Thelaya, K.A., Abdallah, M., Schneider, J., Hamdi, M.: A survey on security and privacy issues in edge computing-assisted internet of things. IEEE Internet Things J. (2020). https://doi.org/10.1109/JIOT.2020.3015432
13. Ogonji, M.M., Okeyo, G., Wafula, J.M.: A survey on privacy and security of internet of things. Comput. Sci. Rev. **38**, 100312 (2020). https://doi.org/10.1016/j.cosrev.2020.100312
14. Fridrich, J.: Steganography in Digital Media: principles, Algorithms, and Applications. Cambridge University Press (2009)
15. Bairagi, A.K., Khondoker, R., Islam, R.: An efficient steganographic approach for protecting communication in the internet of things (IoT) critical infrastructures. Inf. Secur. J.: Glob. Perspect. **25**, 197–212 (2016). https://doi.org/10.1080/19393555.2016.1206640
16. Li, H., Hu, L., Chu, J., Chi, L., Li, H.: The maximum matching degree sifting algorithm for steganography pretreatment applied to IoT. Multimed. Tools Appl. **77**, 18203–18221 (2018). https://doi.org/10.1007/s11042-017-5075-1
17. Devi, S., Sahoo, M.N., Muhammad, K., Ding, W., Bakshi, S.: Hiding medical information in brain MR images without affecting accuracy of classifying pathological brain. Future Gener. Comput. Syst. **99**, 235–246 (2019). https://doi.org/10.1016/j.future.2019.01.047
18. Parah, S.A., Sheikh, J.A., Ahad, F., Bhat, G.M.: High capacity and secure electronic patient record (epr) embedding in color images for IoT driven healthcare systems. In: Internet of Things and Big Data Analytics Toward Next-Generation Intelligence. pp. 409–437. Springer (2018). https://doi.org/10.1007/978-3-319-60435-0_17
19. Parah, S.A., Sheikh, J.A., Akhoon, J.A., Loan, N.A.: Electronic health record hiding in images for smart city applications: a computationally efficient and reversible information hiding technique for secure communication. Future Gener. Comput. Syst. **108**, 935–949 (2020). https://doi.org/10.1016/j.future.2018.02.023
20. Kaw, J.A., Loan, N.A., Parah, S.A., Muhammad, K., Sheikh, J.A., Bhat, G.M.: A reversible and secure patient information hiding system for IoT driven e-health. Int. J. Inf. Manag. **45**, 262–275 (2019). https://doi.org/10.1016/j.ijinfomgt.2018.09.008
21. Horng, J.-H., Xu, S., Chang, C.-C., Chang, C.-C.: An Efficient data-hiding scheme based on multidimensional mini-SuDoKu. Sensors **20**, 2739 (2020). https://doi.org/10.3390/s20092739
22. Huang, C.-T., Tsai, M.-Y., Lin, L.-C., Wang, W.-J., Wang, S.-J.: VQ-based data hiding in IoT networks using two-level encoding with adaptive pixel replacements. J Supercomput. **74**, 4295–4314 (2018). https://doi.org/10.1007/s11227-016-1874-9
23. Sukumar, A., Subramaniyaswamy, V., Vijayakumar, V., Ravi, L.: A secure multimedia steganography scheme using hybrid transform and support vector machine for cloud-based storage. Multimed. Tools Appl. **79**, 10825–10849 (2020). https://doi.org/10.1007/s11042-019-08476-2
24. Cui, Q., Zhou, Z., Fu, Z., Meng, R., Sun, X., Wu, Q.M.J.: Image steganography based on foreground object generation by generative adversarial networks in mobile edge computing with internet of things. IEEE Access **7**, 90815–90824 (2019). https://doi.org/10.1109/ACCESS.2019.2913895
25. Meng, R., Cui, Q., Zhou, Z., Fu, Z., Sun, X.: A Steganography algorithm based on cyclegan for covert communication in the internet of things. IEEE Access **7**, 90574–90584 (2019). https://doi.org/10.1109/ACCESS.2019.2920956
26. Ding, X., Xie, Y., Li, P., Cui, M., Chen, J.: Image Steganography based on artificial immune in mobile edge computing with internet of things. IEEE Access **8**, 136186–136197 (2020). https://doi.org/10.1109/ACCESS.2020.3010513
27. Elhoseny, M., Ramírez-González, G., Abu-Elnasr, O.M., Shawkat, S.A., Arunkumar, N., Farouk, A.: Secure medical data transmission model for IoT-based healthcare systems. IEEE Access **6**, 20596–20608 (2018). https://doi.org/10.1109/ACCESS.2018.2817615

28. Khari, M., Garg, A.K., Gandomi, A.H., Gupta, R., Patan, R., Balusamy, B.: Securing data in internet of things (IoT) using cryptography and steganography techniques. IEEE Trans. Syst. Man Cybern.: Syst. **50**, 73–80 (2020). https://doi.org/10.1109/TSMC.2019.2903785

29. Yassin, A.A., Rashid, A.M., Abduljabbar, Z.A., Alasadi, H.A.A., Aldarwish, A.J.Y.: Toward for strong authentication code in cloud of internet of things based on DWT and steganography. J. Theor. Appl. Inf. Technol. **96**, 2922–2935 (2018)

30. Yan, X., Wang, S., El-Latif, A.A.A., Sang, J., Niu, X.: A novel perceptual secret sharing scheme. In: Shi, Y.Q., Liu, F., Yan, W. (eds.) Transactions on Data Hiding and Multimedia Security IX. Lecture Notes in Computer Science, vol. 8363. Springer, Berlin, Heidelberg (2014). https://doi.org/10.1007/978-3-642-55046-1_5

31. Yan, X., Wang, S., El-Latif, A.A.A., Niu, X.: New approaches for efficient information hiding-based secret image sharing schemes. SIViP **9**, 499–510 (2015). https://doi.org/10.1007/s11760-013-0465-y

32. Jiang, S., Ye, D., Huang, J., Shang, Y., Zheng, Z.: SmartSteganogaphy: light-weight generative audio steganography model for smart embedding application. J. Netw. Comput. Appl. **165**, 102689 (2020). https://doi.org/10.1016/j.jnca.2020.102689

33. Anguraj, S., Shantharajah, S.P., Emilyn, J.J.: A steganographic method based on optimized audio embedding technique for secure data communication in the internet of things. Comput. Intell. **36**, 557–573 (2020). https://doi.org/10.1111/coin.12253

34. Pizzolante, R., Castiglione, A., Carpentieri, B., De Santis, A., Palmieri, F., Castiglione, A.: On the protection of consumer genomic data in the internet of living things. Comput. Secur. **74**, 384–400 (2018). https://doi.org/10.1016/j.cose.2017.06.003

35. Hurrah, N.N., Parah, S.A., Sheikh, J.A., Al-Turjman, F., Muhammad, K.: Secure data transmission framework for confidentiality in IoTs. Ad Hoc Netw. **95**, 101989 (2019). https://doi.org/10.1016/j.adhoc.2019.101989

36. Al-Shayea, T.K., Batalla, J.M., Mavromoustakis, C.X., Mastorakis, G.: Embedded dynamic modification for efficient watermarking using different medical inputs in IoT. In: 2019 IEEE 24th International Workshop on Computer Aided Modeling and Design of Communication Links and Networks (CAMAD). pp. 1–6 (2019). https://doi.org/10.1109/CAMAD.2019.8858489

37. Al-Shayea, T.K., Mavromoustakis, C.X., Batalla, J.M., Mastorakis, G., Pallis, E., Markakis, E.K., Panagiotakis, S., Khan, I.: Medical image watermarking in four levels decomposition of DWT using multiple wavelets in IoT emergence. In: Mastorakis, G., Mavromoustakis, C.X., Batalla, J.M., and Pallis, E. (eds.) Convergence of Artificial Intelligence and the Internet of Things. pp. 15–31. Springer International Publishing, Cham (2020). https://doi.org/10.1007/978-3-030-44907-0_2

38. Huo, Y., Liu, J., Zhang, Q., Zeng, Y., Yang, F., Chang, J., Fan, Y., Liu, C.: Semi-fragile watermarking for color image authentication in power internet of things. In: 2019 IEEE Innovative Smart Grid Technologies—Asia (ISGT Asia). pp. 2865–2869 (2019). https://doi.org/10.1109/ISGT-Asia.2019.8881663

39. Lin, Y.-H., Hsia, C.-H., Chen, B.-Y., Chen, Y.-Y.: Visual IoT security: data hiding in AMBTC images using block-wise embedding strategy. Sensors **19**, 1974 (2019). https://doi.org/10.3390/s19091974

40. Liu, J., Ma, J., Li, J., Huang, M., Sadiq, N., Ai, Y.: Robust watermarking algorithm for medical volume data in internet of medical things. IEEE Access **8**, 93939–93961 (2020). https://doi.org/10.1109/ACCESS.2020.2995015

41. Benrhouma, O., Hermassi, H., El-Latif, A.A.A., Belghith, S.: Chaotic watermark for blind forgery detection in images. Multimed. Tools Appl. **75**, 8695–8718 (2016). https://doi.org/10.1007/s11042-015-2786-z

42. Wang, J., Smith, G.L.: A cross-layer authentication design for secure video transportation in wireless sensor network. IJSN **5**, 63 (2010). https://doi.org/10.1504/IJSN.2010.030724

43. Tondwalkar, A., Vinayakray-Jani, P.: Secure localisation of wireless devices with application to sensor networks using steganography. Procedia Comput. Sci. **78**, 610–616 (2016). https://doi.org/10.1016/j.procs.2016.02.107

44. Kim, S.R., Kim, J.N., Kim, S.T., Shin, S., Yi, J.H.: Anti-reversible dynamic tamper detection scheme using distributed image steganography for IoT applications. J Supercomput. **74**, 4261–4280 (2018). https://doi.org/10.1007/s11227-016-1848-y

45. Das, R., Das, I.: Secure data transfer in IoT environment: adopting both cryptography and steganography techniques. In: 2016 Second International Conference on Research in Computational Intelligence and Communication Networks (ICRCICN), pp. 296–301 (2016). https://doi.org/10.1109/ICRCICN.2016.7813674

46. Bapat, C., Baleri, G., Inamdar, S., Nimkar, A.V.: Smart-lock security re-engineered using cryptography and steganography. In: Thampi, S.M., Martínez Pérez, G., Westphall, C.B., Hu, J., Fan, C.I., and Gómez Mármol, F. (eds.) Security in Computing and Communications, pp. 325–336. Springer, Singapore (2017). https://doi.org/10.1007/978-981-10-6898-0_27

47. Zaichenko, D.S., Sineva, I.S.: The Study of Genetic type steganographic models to increase noise immunity of IoT systems. Int. J. Embed. Real-Time Commun. Syst. (IJERTCS) **11**, 1–15 (2020). https://doi.org/10.4018/IJERTCS.2020040101

48. Yang, W., Wang, S., Hu, J., Ibrahim, A., Zheng, G., Macedo, M.J., Johnstone, M.N., Valli, C.: A cancelable iris- and steganography-based user authentication system for the internet of things. Sensors **19**, 2985 (2019). https://doi.org/10.3390/s19132985

49. Wang, H.: Communication-resource-aware adaptive watermarking for multimedia authentication in wireless multimedia sensor networks. J Supercomput. **64**, 883–897 (2013). https://doi.org/10.1007/s11227-010-0500-5

50. Peng, H., Yang, B., Li, L., Yang, Y.: Secure and traceable image transmission scheme based on semitensor product compressed sensing in telemedicine system. IEEE Internet Things J. **7**, 2432–2451 (2020). https://doi.org/10.1109/JIOT.2019.2957747

51. Yoon-Su, J., Seung-Soo, S.: Staganography-based healthcare model for safe handling of multimedia health care information using VR. Multimed. Tools Appl. **79**, 16593–16607 (2020). https://doi.org/10.1007/s11042-019-07833-5

52. Khosravi, M.R., Samadi, S.: Efficient payload communications for IoT-enabled ViSAR vehicles using discrete cosine transform-based quasi-sparse bit injection. J. Wirel. Com. Netw. **2019**, 262 (2019). https://doi.org/10.1186/s13638-019-1572-4

53. Janakiraman, S., Thenmozhi, K., Rayappan, J.B.B., Amirtharajan, R.: Indicator-based lightweight steganography on 32-bit RISC architectures for IoT security. Multimed. Tools Appl. **78**, 31485–31513 (2019). https://doi.org/10.1007/s11042-019-07960-z

54. Li, L., Hossain, M.S., El-Latif, A.A.A., Alhamid, M.F.: Distortion less secret image sharing scheme for internet of things system. Clust. Comput. **22**, 2293–2307 (2019). https://doi.org/10.1007/s10586-017-1345-y

55. Pu, Y.-F., Zhang, N., Wang, H.: Fractional-order spatial steganography and blind steganalysis for printed matter: anti-counterfeiting for product external packing in internet-of-things. IEEE Internet Things J. **6**, 6368–6383 (2019). https://doi.org/10.1109/JIOT.2018.2886996

56. Abd-El-Atty, B., Iliyasu, A.M., Alaskar, H., El-Latif, A.A.A.: A robust quasi-quantum walks-based steganography protocol for secure transmission of images on cloud-based E-healthcare platforms. Sensors **20**, 3108 (2020). https://doi.org/10.3390/s20113108

57. El-Latif, A.A.A., Abd-El-Atty, B., Elseuofi, S., Khalifa, H.S., Alghamdi, A.S., Polat, K., Amin, M.: Secret images transfer in cloud system based on investigating quantum walks in steganography approaches. Phys. A: Stat. Mech. Appl. **541**, 123687 (2020). https://doi.org/10.1016/j.physa.2019.123687

58. Peng, J., El-Atty, B.A., Khalifa, H.S., El-Latif, A.A.A.: Image steganography algorithm based on key matrix generated by quantum walks. In: Eleventh International Conference on Digital Image Processing (ICDIP 2019). p. 1117905 (2019). https://doi.org/10.1117/12.2539630

59. EL-Latif, A.A.A., Abd-El-Atty, B., Venegas-Andraca, S.E.: A novel image steganography technique based on quantum substitution boxes. Opt. Laser Technol. **116**, 92–102 (2019). https://doi.org/10.1016/j.optlastec.2019.03.005

60. de Fuentes, J.M., Blasco, J., González-Tablas, A.I., González-Manzano, L.: Applying information hiding in VANETs to covertly report misbehaving vehicles. Int. J. Distrib. Sens. Netw. **10**, 120626 (2014). https://doi.org/10.1155/2014/120626

61. Ren, J., Wu, G., Yao, L.: A sensitive data aggregation scheme for body sensor networks based on data hiding. Pers Ubiquit Comput. **17**, 1317–1329 (2013). https://doi.org/10.1007/s00779-012-0566-6

62. Sahu, N., Peng, D., Sharif, H.: An innovative approach to integrate unequal protection-based steganography and progressive transmission of physiological data. SN Appl. Sci. **2**, 237 (2020). https://doi.org/10.1007/s42452-020-1992-0

63. Hussain, I., Negi, M.C., Pandey, N.: Security in ZigBee using steganography for IoT communications. In: Kapur, P.K., Klochkov, Y., Verma, A.K., and Singh, G. (eds.) System Performance and Management Analytics. pp. 217–227. Springer, Singapore (2019). https://doi.org/10.1007/978-981-10-7323-6_18

64. Zhang, W., Liu, Y., Das, S.K., De, P.: Secure data aggregation in wireless sensor networks: a watermark based authentication supportive approach. Pervasive Mob. Comput. **4**, 658–680 (2008). https://doi.org/10.1016/j.pmcj.2008.05.005

65. Panah, A.S., van Schyndel, R., Sellis, T., Bertino, E.: In the shadows we trust: a secure aggregation tolerant watermark for data streams. In: 2015 IEEE 16th International Symposium on a World of Wireless, Mobile and Multimedia Networks (WoWMoM). pp. 1–9 (2015). https://doi.org/10.1109/WoWMoM.2015.7158149

66. Yavari, A., Panah, A.S., Georgakopoulos, D., Jayaraman, P.P., van Schyndel, R.: Scalable role-based data disclosure control for the internet of things. In: 2017 IEEE 37th International Conference on Distributed Computing Systems (ICDCS). pp. 2226–2233 (2017). https://doi.org/10.1109/ICDCS.2017.307

67. Yue, M., Peng, Z., Zheng, K., Peng, Y.: Rights protection for trajectory streams. In: Database Systems for Advanced Applications, pp. 407–421. Springer, Cham (2014). https://doi.org/10.1007/978-3-319-05813-9_27

68. Lipuš, B., Žalik, B.: Robust watermarking of airborne LiDAR data. Multimed. Tools Appl. **77**, 29077–29097 (2018). https://doi.org/10.1007/s11042-018-6039-9

69. Chong, S., Skalka, C., Vaughan, J.A.: Self-identifying data for fair use. J. Data Inf. Quality. **5**(11), 1–11, 30 (2015). https://doi.org/10.1145/2687422

70. Hoang, T.-M., Bui, V.-H., Vu, N.-L., Hoang, D.-H.: A lightweight mixed secure scheme based on the watermarking technique for hierarchy wireless sensor networks. In: 2020 International Conference on Information Networking (ICOIN), pp. 649–653 (2020). https://doi.org/10.1109/ICOIN48656.2020.9016541

71. Xiao, Y., Gao, G.: Digital watermark-based independent individual certification scheme in WSNs. IEEE Access **7**, 145516–145523 (2019). https://doi.org/10.1109/ACCESS.2019.2945177

72. Wang, C., Bai, Y., Mo, X.: Data secure transmission model based on compressed sensing and digital watermarking technology. Wuhan Univ. J. Nat. Sci. **19**, 505–511 (2014). https://doi.org/10.1007/s11859-014-1045-x

73. Chen, T.-S., Hou, K.-N., Beh, W.-K., Wu, A.-Y.: Low-complexity compressed-sensing-based watermark cryptosystem and circuits implementation for wireless sensor networks. IEEE Trans. Very Large Scale Integr. (VLSI) Syst. **27**, 2485–2497 (2019). https://doi.org/10.1109/TVLSI.2019.2933722

74. Kamel, I., Juma, H.: Simplified watermarking scheme for sensor networks. IJIPT **5**, 101 (2010). https://doi.org/10.1504/IJIPT.2010.032619

75. Shi, X., Xiao, D.: A reversible watermarking authentication scheme for wireless sensor networks. Inf. Sci. **240**, 173–183 (2013). https://doi.org/10.1016/j.ins.2013.03.031

76. Wang, B., Kong, W., Li and Neal N. Xiong, W.: A dual-chaining watermark scheme for data integrity protection in internet of things. Comput. Mater. Continua. **58**, 679–695 (2019). https://doi.org/10.32604/cmc.2019.06106

77. Ferdowsi, A., Saad, W.: Deep learning for signal authentication and security in massive internet-of-things systems. IEEE Trans. Commun. **67**, 1371–1387 (2019). https://doi.org/10.1109/TCOMM.2018.2878025

78. Hameed, K., Khan, A., Ahmed, M., Goutham Reddy, A., Rathore, M.M.: Towards a formally verified zero watermarking scheme for data integrity in the internet of things based-wireless

sensor networks. Future Gener. Comput. Syst. **82**, 274–289 (2018). https://doi.org/10.1016/j.future.2017.12.009

79. Nguyen, V.-T., Hoang, T.-M., Duong, T.-A., Nguyen, Q.-S., Bui, V.-H.: A lightweight watermark scheme utilizing MAC layer behaviors for wireless sensor networks. In: 2019 3rd International Conference on Recent Advances in Signal Processing, Telecommunications Computing (SigTelCom), pp. 176–180 (2019). https://doi.org/10.1109/SIGTELCOM.2019.8696234

80. Xu, Z., Yuan, W.: Watermark BER and channel capacity analysis for QPSK-based RF watermarking by constellation dithering in AWGN channel. IEEE Signal Process. Lett. **24**, 1068–1072 (2017). https://doi.org/10.1109/LSP.2017.2710144

81. Huang, H., Zhang, L.: Reliable and secure constellation shifting aided differential radio frequency watermark design for NB-IoT systems. IEEE Commun. Lett. **23**, 2262–2265 (2019). https://doi.org/10.1109/LCOMM.2019.2944811

82. Mo, Y., Weerakkody, S., Sinopoli, B.: Physical authentication of control systems: designing watermarked control inputs to detect counterfeit sensor outputs. IEEE Control Syst. Mag. **35**, 93–109 (2015). https://doi.org/10.1109/MCS.2014.2364724

83. Rubio-Hernan, J., De Cicco, L., Garcia-Alfaro, J.: Adaptive control-theoretic detection of integrity attacks against cyber-physical industrial systems. Trans. Emerg. Telecommun. Technol. **29**, e3209 (2018). https://doi.org/10.1002/ett.3209

84. Song, Z., Skuric, A., Ji, K.: A recursive watermark method for hard real-time industrial control system cyber-resilience enhancement. IEEE Trans. Autom. Sci. Eng. **17**, 1030–1043 (2020). https://doi.org/10.1109/TASE.2019.2963257

Survey on Mobile Edge-Cloud Computing: A Taxonomy on Computation offloading Approaches

Ibrahim A. Elgendy and Rahul Yadav

Abstract With the technological evolution of Internet of Things (IoT) devices and wireless communications, a wide variety of new complex mobile applications and different services have rapidly increased. Nevertheless, these devices are considered constrained to processing such applications, due to the limitation of battery capacity and high-demand computation for these applications. Mobile cloud comping (MCC) is considered as an appropriate solution for addressing this problem and battery the battery lifetime of these devices, in which the intensive-computation tasks will be offloaded and processed at a conventional centralized cloud. However, cloud computing solution introduces a high communication delay which makes the computation offloading inappropriate for processing real-time applications. To tackle the problem of delay, a new emerging paradigm has been introduced, called mobile edge computing (MEC), in which the computation and storage capabilities of cloud computing have been provided at the edge of the network that enables such applications to be processed as well as satisfying the delay requirements. To this end, compared to other surveys, this paper provides a comprehensive survey of state-of-the-art MEC research with a focus on computation offloading on edge-cloud computing combination. In addition, we provide a novel taxonomy on computation offloading at edge-cloud computing combination and introduce the most and common recent computation offloading models regarding this taxonomy. Furthermore, we highlight the main strengths, weaknesses and other issues which require further consideration. Finally, open research challenges and new research trends in edge-cloud computing will be discussed.

I. A. Elgendy (✉)
School of Computer Science and Technology, Harbin Institute of Technology, P. Box 75, Harbin, China
e-mail: ibrahim.elgendy@hit.edu.cn

Department of Computer Science, Faculty of Computers and Information, Menoufia University, Shibin el Kom 32511, Egypt

R. Yadav
Peng Cheng Laboratory, Shenzhen, China
e-mail: rahulyd@pcl.ac.cn

1 Introduction

The mobile Internet of Things (IoT) devices and wireless sensors are becoming more popular and playing increasingly important roles in every aspect of our daily life [2, 4, 53, 119]. In addition, it was expected that sensor-enabled objects connected to the network would rise to more than 75 billion by 2025. Moreover, with the availability of high-speed and stable internet, new technologies and a broad range of multi-functional sensors and applications have been birthed which are continuously generating a massive amount of data that requires to be processed in a real-time such as augmented and virtual reality, facial recognition, video games, e-health, social networks, transportation and natural language processing [7, 25, 51]. Most of these applications require an exponential growth of computation and high data rate beyond the capacity of embedded IoT devices. However, the limited computation power and energy remain an essential obstacle to the completion of these data locally in real-time by these devices [125, 139].

In general, to save energy and extending battery lifetime of these devices, major hardware and software level changes are needed which can be classified into four basic approaches [69]:

- *Adopt a new generation of semiconductor technology*: Although transistors are smaller and consume less power, the battery needs more transistors for supporting better performance. Therefore, power consumption increases.
- *Avoid wasting energy*: In this approach, system's components go sleep to save energy like dim the display while not using.
- *Execute programs slowly*: When a processor's clock speed doubles, the power consumption nearly couplets. If half reduces the clock speed, the execution time doubles, but only one-quarter of the energy is consumed.
- *Computation Offloading*: The mobile system does not perform the computation; instead, it offloads intensive methods of mobile applications to run remotely on a rich resource such as cloud or edge, thereby extending the mobile system's battery lifetime, as shown in Fig. 1.

This paper will focus on the last approach (i.e., Computation Offloading), in which the intensive tasks will be offloaded and executed at richer resources to save energy and extend the lifetime of battery. Mobile cloud computing is one of the available paradigms that can mitigate the limitations of these devices through the concept of computation offloading. Specifically, the intensive tasks would be transmitted and remotely processed at the centralized cloud through wireless channels [35, 98]. However, the high latency associated with continuous or tightly coupled communication with cloud services is a significant drawback that makes mobile cloud computing undesirable paradigm for real-time and delay-sensitive applications [16, 42, 43, 66, 97]. Moreover, IoT devices are susceptible to several threats of security during the different phases of data transmission [5, 6, 8, 73, 91, 92, 120, 147].

To address these challenges, researchers have realized that the most cost-effective and low-latency solution is utilizing the resources and services that is closer to IoT

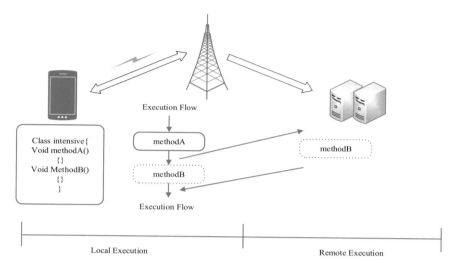

Fig. 1 Computational offloading

mobile networks. This led to a new paradigm namely edge computing [67, 87, 130]. Cloudlet is the first paradigm applying the edge computing concept that proposed in 2009, in which the computing/storage is moved closer to IoT devices [39, 113]. The cloudlet's main idea is to put servers with high processing capacity at strategic positions to deliver both the computational and storage resources for the surrounding IoT devices which is similar to Wi-Fi hotspots scenario, except the cloudlet provides cloud services for IoT devices instead of internet access. However, cloudlet is not considered as a suitable solution for the large-scale mobile IoT networks. In addition, it is difficult to fulfill Quality of Service (QoS) for IoT devices, much like MCC. Furthermore, Wi-Fi is utilized by cloudlet to coverage IoT devices, which is only available locally with limited support of mobility. The other alternative solution that provides the cloud computing capabilities at the edge is ad-hoc network paradigm, in which set of IoT devices in proximity are combined their computation resources to process the high demanding applications locally [13, 28, 41, 140]. However, this solution poses set of critical challenges that need to be handled such as (1) Discovering the proper proximity of IoT networks to provide the computing recourses and to guarantee that the processed data are returned to the source of IoT device. (2) Coordination between the computing resources of IoT networks has to be enabled despite the fact that there are no control channels to facilitate reliable computing. (3) Security and privacy issues[37].

From the point of view of mobile IoT networks, high latency, traffic overhead, security [3, 34, 81, 144] and privacy problems are the most prominent drawbacks for the above-mentioned edge computing principles. Therefore, moving the cloud computing capabilities at edge nodes (Base stations) which is closer to mobile devices as well as connected to the cloud servers via core network be the most suitable solution for mobile IoT networks, called mobile edge-cloud computing [78, 84], as shown in

Fig. 2. In recent years, mobile edge-cloud computing has gained much attention and popularity in which it provides storage, computing and services at the edge of the network or via a radio access network. Moreover, it facilitates to reduce the network latency, improve the quality of services and reduce the security threats in comparison with traditional cloud computing paradigm [10, 36, 44–46, 101].

Recently, there are several survey papers on mobile cloud computing and MEC have been published, shown in Table 1. For mobile cloud computing, Sanaei et al. [112] surveyed the heterogeneity problem among mobile devices and cloud computing vendors and also discussed the related challenges. Additionally, Kovachev et al. [68] and Khan et al. [64] presented level comparison of mobile cloud computing application models and highlighted their advantages and disadvantages. Furthrmore, in [32, 50], the authors addressed the generic issues of the mobile cloud computing. Whereas for MEC, Wang et al. [126, 133] surveyed the mobile edge networks in which the integration between computing, caching and communications are addressed. In [29, 110, 136], the authors extensively analyzed the security threats and its challenges for MEC. In [23, 71, 134, 145], the authors surveyed the convergence of deep learning and edge computing. In [48], the authors handled the mobility issue and how to mange it for IoT devices based on IP protocol. In [9, 61, 76, 84],

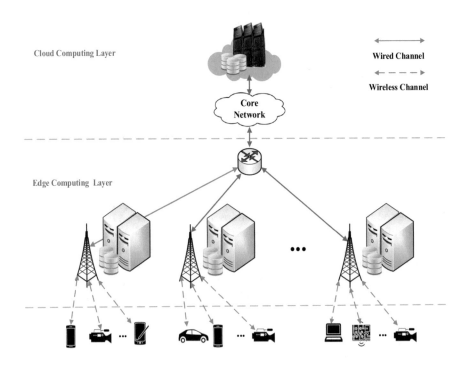

Fig. 2 Edge computing architecture

Table 1 Comparison of survives on MCC and MEC

Ref.	Published in	Year	Area focused
Chen et al. [23]	Proceedings of the IEEE	2019	Deep Learning and Edge Computing Integration
Dinh et al. [32]	Wireless communications and mobile computing	2013	Architecture, applications, and approaches of MCC
Ghaleb et al. [48]	Journal on Wireless Communications and Networking	2016	Mobility Management for IoT
Guan et al. [50]	IEEE/ACIS	2011	Research Effort Towards MCC
Wang et al. [134]	IEEE Communications Surveys & Tutorials	2020	Convergence of Edge Computing and Deep Learning
Jiang et al. [61]	IEEE Access	2019	Various Aspects of Computation Offloading in MEC
Khan et al. [64]	IEEE Communications Surveys & Tutorials	2013	Application Models of MCC
Kovachev et al. [68]	ArXiv	2011	Application Models of MCC
Li et al. [71]	IEEE Network	2018	Deep Learning for IoTs into the EC
Lin et al. [76]	Proceedings of the IEEE	2019	Computation Offloading Approaches For MEC
Mach et al. [84]	IEEE Communications Surveys & Tutorials	2017	Architecture and Computation Offloading of MEC
Roman et al. [110]	Future Generation Computer Systems	2018	Security Issues in Foge and Edge Computing
Sanaei et al. [112]	IEEE Communications Surveys & Tutorials	2013	Taxonomy and Open challenges for MCC
Wang et al. [126]	IEEE Communications Surveys & Tutorials	2017	Networking, Caching, and Computing integration in Wireless Systems
Wang et al. [133]	IEEE Access	2017	Convergence of Computing, Caching and Communications for MEC
Xiao et al. [136]	Proceedings of the IEEE	2019	Security Issues in Edge Computing
Zhang et al. [145]	IEEE Communications Surveys & Tutorials	2019	Deep Learning in Mobile and Wireless Networking
Cui et al. [29]	Security and Communication Networks	2021	Secure Deployment of Mobile Services in Edge Computing
Afrin et al. [9]	IEEE Communications Surveys & Tutorials	2021	Resource Allocation and Service Provisioning in Multi-Agent Cloud Robotics
Our Paper	–	2021	Taxonomy of Computation Offloading Models on Edge-Cloud Computing

the authors surveyed the computation offloading models for MEC. However, most of the above-mentioned surveys do not make any thematic taxonomy for the current computation offloading models for edge-cloud computing. Therefore, in this paper, we provide a systematic taxonomy for the current computational offloading models for mobile edge-cloud computing to investigate and categorize related research articles. Additionally, an overview of the common models based on this taxonomy and their related challenges will be discussed.

The rest of this paper is organized as follows. Section 2 introduces the background about edge-cloud computing. Following that, Sect. 3 presents a new comprehensive taxonomy of computation offloading models on edge-cloud computing. The main challenges and the critical open issues for future work are presented in Sect. 4. Finally, Sect. 5 concludes the paper.

2   Edge Computing Background

2.1   Characteristics of Edge computing

Edge computing pushes the cloud computing capabilities at edge nodes which are place in proximity to IoT devices. In addition, it is characterized by set of features which are on-premises, proximity, lower latency, e-location awareness, and network context information, as shown in Fig. 3. These characteristics are briefly described in the following.

- *On-Premises*: The computing resources are located at customer premises which are isolated from the rest of network and can be managed by network operator. In addition, it retains sensitive data on-premises while still taking advantage of the edge cloud's elasticity. Furthermore, the segregation property makes edge computing less vulnerable to attacks.

Fig. 3 Edge computing characteristics

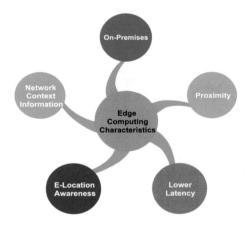

- *Proximity*: Edge computing resources are deployed at nearest location which can process, analyze and materialize big data [95]. Moreover, it is also beneficial for processing the resource-demanding applications such as augmented reality, video analytics, smart warehouse, health-care, etc. as we will discuss later.
- *Lower Latency*: Mobile service's latency is the summation of transmission, computation, and receiving time which depends on the computation capacity and data rate, respectively [47, 115]. For the edge computing, the mobile services are deployed at the nearest location to IoT devices which led to provide high quality with ultra-low latency and high bandwidth.
- *E-Location Awareness*: Edge computing is aware of all the edge device where it receives the location information from each IoT device that exists within the local access network. Then, this information can be used for making a smart decision.
- *Network Context Information*: Edge computing server has the ability to leverage from the proximity of IoT devices by tracking their real-time information such as their locations, behaviors, environments and context. Based on this information, edge computing server can estimate the radio cell congestion which further helps them to make smart decisions for improving the delivery of context-aware services to the IoT devices. For example, in the museum video guide, the augmented reality application can predict the interests of the users on the basis of their location and then provides its related contents such as antiques and artworks [11, 114].

2.2 Benefits of Edge Computing

Edge computing brings several advantages and benefits in comparison with its counterparts such as cloud computing and cloudlets such as it increases the speed of service delivery, enhances the security and privacy, scalability, reliability, versatility, and Energy Saving, as shown in Fig. 4. These benefits are briefly described with some examples in the following. These characteristics are briefly described in the following.

- *Speed*: Speed is one of the most important benefits of edge computing where it has the ability to increase the performance of the network and process the data of real-time application with low latency. Since, the application's data of IoT devices can be processed locally or in nearby edge node instantaneously which has large expansive implications for critical applications such as health-care systems and self-driving cars [107, 154].
- *Security/Privacy Enhancement*: Another key feature of edge computing is the capability of enhancing the mobile applications' security and privacy in comparison with cloud computing where the application's data are processed across a wide range of edge nodes which reduces the probability to become the target of a security attack as well as makes it easier for implementing different security protocols which can seal off compromised portions without shutting down the entire network. Moreover, edge computing can process most of the application's data on

Fig. 4 Edge computing
benefits

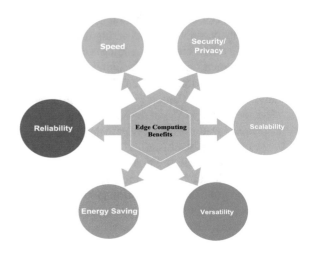

local devices instead of transmitting it to a central data center and can manage the
authorization, access control, and classify service requests' without any third party
[110, 141]. For example, in the health-care system, wearable devices record the
patients' information periodically which is considered as a personal information
and should be saved carefully in private place to prevent their leakage. So, by using
edge computing, this information will be saved locally and the privacy problem
will be overcome.

- *Scalability*: Edge computing offers the ability of scaling the computing and storage
  resources up and down with less expensive costs which allow the companies for
  expanding the computing capacity according to theirs needs easily, quickly and
  cost-effectively. In addition, the scalability benefit does not impose substantial
  bandwidth demands on the core of network.
- *Versatility*: The scalability feature of edge computing boosts the versatility by
  helping the companies to join with local edge data centers and reach the target
  desirable markets easily and without requiring costly infrastructure expansions.
- *Reliability*: Edge computing can provide IoT devices with an effective and unpar-
  alleled reliable resources and services due to it's computing servers were placed
  closer to end users which allows them to process the vital processing functions
  natively. In addition, the shutdown service due to one device failure is entirely
  difficult to occur because there are many edge computing servers and data centers
  are connected through network.
- *Energy Saving*: As mentioned above, most of IoT devices have limited computation
  power and energy which restricts them for processing large data and complex
  application such as augmented reality, video games, health-care and surveillance.
  Hence, edge computing is considered one of promising solution which can provide
  these devices with computation capabilities and storage through effectively using
  computation offloading. More specifically, computation-intensive tasks can be

offloaded and processed at the edge nodes for minimizing the energy consumption and extending the battery lives of these devices [38, 87].

2.3 What to be Processed Through Edge Computing?

Recently, we have seen the explosion growth in the number of IoT devices where it was increased from 500 million to 25 billion devices over the last 10 years, but it will be expected to reach to a trillion devices over the next 10 years [20]. In addition, these devices generate enormous volumes of data continuously that needs to be analyzed and processed quickly. Edge computing is considered a prominent solution the has the ability to process and analyze the data of any IoT device in a real-time by providing a computational power and storage in the form of local computing power and storage without needing to cloud or data centers transmissions. Furthermore, the type and delivery of content of the IoT applications are improved.

There are many use cases and different tasks that already deployed based on edge computing paradigm which can benefit from its features such as augmented reality and virtual reality which are benefit from low-latency communications and quick response, industrial IoT applications that depend on smart network resources utilization, autonomous vehicles which also require high-bandwidth, high availability settings and low-latency, financial application also adopting edge computing to provide a better target services to customers, health-care system benefit from integrating edge computing which offers a new exciting possibilities for delivering patient care and others examples as we will show later in details [105, 123].

2.4 Edge Computing Application Areas

There are different scenarios and applications that need to analyze and process big data in rear-time, but cannot leverage from using the centralized cloud computing due to the bandwidth and connection problem. Hence, processing this data at the edge computing is considered the right decision which can bring a new value to the application and organization [102]. Autonomous vehicle, virtualized radio networks and 5G, video surveillance and analytics, smart building, smart Warehousing, smart manufacturing, smart trafic, smart transportation systems, industrial automation, augmented reality (AR) and virtual reality (VR), connected homes and offices, power plants, factories, ports, medical/health-care systems, cyber-physical systems and retail industry are considered some examples of hundreds or perhaps thousands of examples where edge computing is being deployed. In the following subsections, we will show how these applications are deployed using edge computing paradigm.

3 Edge-Cloud Computing Taxonomy

Recently, various optimization models have been developed to raise the pitfalls faced by IoT devices. In this section, we provide a systematic taxonomy for the current computational offloading models for mobile edge-cloud computing which is shown in Fig. 5. Additionally, an overview of the common models based on this taxonomy and their related challenges will be discussed in the next subsections.

3.1 Models Objective

Regarding the first category of the taxonomy, the model objectives represents the objectives that must be achieved for making the offloading decision such as minimizing execution delay, minimizing energy consumption, minimizing overall cost, maximize throughput, minimize radio utilization, user preference, and/or optimizing the computing resources. We can classify the current models into three main categories which are mono-objective, bi-objective and multi-objective. Mono-objective model indicates that the user equipment will migrate or offload the application's computations to achieve only single objective from the mentioned objectives. While in bi-objective model, the user equipment will achieve only two objectives for applying the offloading. But finally, in the multi-objective model, multiple objectives can be fulfilled for making the offloading decision.

3.1.1 Mono-Objective

In this subsection, minimizing the energy consumption or reducing the delay of execution at IoT devices are the key objective of the surveyed papers.

First, Zhang et al. [146] proposed an energy-efficient computation offloading (EECO) framework for multi-user mobile edge cloud computing in fifth-generation heterogeneous networks. Additionally, computation task offloading and radio resource allocation are jointly formulated as optimization problem whose objective is to minimize the energy consumption of the entire system. Furthermore, a three-stage EECO algorithm was designed to cope with system complexity of this problem and find the computation offloading decision in an efficient manner. $\mathcal{O}(\max(I2 + N, IK + N))$ is the time complexity for this algorithm in which I, N and K denote the number of iterations, the number of mobile devices and the number of available channels. The experiential results verified that the proposed framework reduces the system consumption of energy by up to 15% in comparison with local execution. Moreover, the proposed framework can optimally decide to offload the computation tasks for the large-scale of mobile devices. However, mobile application is treated as a single unit which will be offloaded or not, thereby consumed

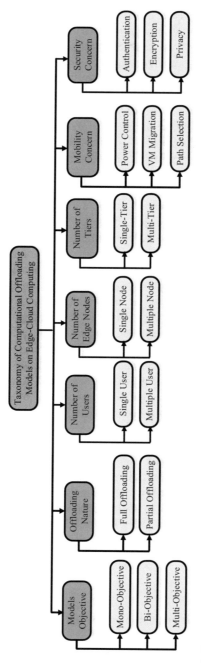

Fig. 5 Taxonomy of computational offloading models

more resources due to a large amount of data will be transmitted via the network. In addition, the application data are not secured during the transmission process.

The same objective of [146] is achieved in [118], in which a set of wireless sensor nodes are cooperated to provide a cooperative computing. In addition, the mobile applications are assumed to be partitioned into independent tasks which will be assigned to one of the peer cooperation nodes for execution where the node selection is based on the tradeoff between fairness and energy consumption. The numerical and simulation results demonstrate that the proposed optimal policy can reduce the system energy consumption by up to 10% in comparison with local execution. Nevertheless, the main drawback of the proposed method is that quality of service and quality of experience (QoE) for users can be hardly guaranteed because it's difficult to identify a set of wireless sensors nodes in proximity and guarantee that they will deliver the processed data back to the source node. In addition, a single mobile device user scenario is only considered in this study.

Secondly, reducing the latency of mobile applications for resource-constrained mobile devices is the main goal of [62, 108]. Specifically, Kao et al. [62] formulated an integer optimization problem which optimizes the computation task assignment in which application-specific profile, link connectivity and availability of computational resources are jointly considered. In addition, a novel fully polynomial-time approximation approach is proposed to solve this problem and find the task assignment solution. Furthermore, an online learning algorithm is developed to guarantee the performance gap in comparison with the optimal strategy. The simulation results proved that the proposed framework reduces the latency by 16% in comparison with heuristic approach. Meanwhile, Ren et al. [108] have jointly considered the computation and communication resources allocations for TDMA-based multiuser MEC system. In addition, three different video compression offloading models, namely local compression, edge cloud compression, and partial compression offloading, are studied in which an optimization problem is formulated whose objective is to reduce the weighted sum delay of the mobile device users. Numerical results proved that partial compression offloading can efficiently reduce the system delay by up to 61% and 45% in comparison with local and edge cloud compression, respectively. However, the authors in [62] did not consider the data traffic optimization that was incurred during task execution as well as did not protect the application data from attacks during transmission to the edge server. In addition, the advanced wireless communication techniques are not considered in [108], in which it can enhance the performance of edge computing systems.

3.1.2 Bi-Objective

This section highlights the common models that deal with only two objectives in computation offloading of the MEC system.

Minimizing the energy consumption and reducing the execution delay of applications for MEC systems is considered the main goal of [38, 63, 90]. In [63], an inter-layer optimization of computation offloading and transmission power over the

Fig. 6 System model of [38]

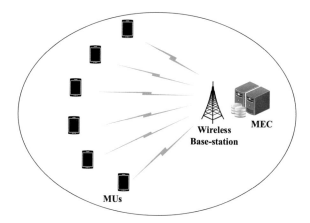

physical layer and the application layer, respectively, are investigated. In addition, to reduce the complexity of computation offloading, an efficient message-passing framework based-on the application's topological structure is proposed which can be applied to both the conventional and parallel implementations of processing and communication. The work in [90] formulated a power consumption minimization problem with the application's task buffer stability constraints. Then, Lyapunov optimization-based online algorithm is developed which optimally decided how to allocate the transmission power and bandwidth for mobile device users' offloading to MEC. The simulation results proved that the energy consumption and execution delay could be minimized by up to 90% and 98%, respectively, by applying the computation offloading on MEC. Whereas in [38], the authors proposed a multi-user computation offloading and resource allocation model for MEC, shown in Fig. 6. In addition, a new security layer based on advanced encryption standard (AES) cryptographic technique with a genetic algorithm is added to protect the application data during the transmission. The simulation results show that the proposed model with and without security addition can respectively save about 17% and 14% energy of the entire system. Nevertheless, in [63], the applications' data does not protect from cyber-attacks during transmission. In addition, the transmission interference between mobile users is not considered in [90] where it has a great effect on the system performance. Moreover, the computation tasks' deadline requirement is not imposed.

3.1.3 Multi-objective

In this section, we will survey the papers that address the multi-objectives computation offloading for mobile edge-cloud computing.

Liu et al. [80] have explicitly and jointly optimized the wireless transmission power and the computation offloading probability through formulating an optimization problem a queuing model whose aim is to minimize the consumption of energy, execution delay and price cost for multi-user mobile edge-cloud computing. In addition, the scalarization approach and interior point method are applied to derive the optimal solution. The main drawback of the proposed method is that the tasks' requirements diversity in the resource management strategy is not considered, in which it will be more significant and effective in the MEC system.

Another idea addressing multi-objective computation offloading is introduced in [137, 149]. When compared to the previous study, Xu et al. [137] proposed a multi-objective computation model for internet of vehicle in edge-cloud computing while minimizing the energy consumption, reducing the execution delay and decreasing the load balancing rate for edge computing devices are the main goal of this model. First, a vehicle-to-vehicle communication-based algorithm is designed to find the offloading route for the computation tasks. In addition, a non-dominated sorting genetic algorithm is adopted for addressing the optimization problem and obtain the solution. Furthermore, simple additive weighting and multiple criteria decision-making methods are employed for evaluating the generated solution. Whereas in [149], Zhang et al. investigated a multi-objective resource allocation for multi-user orthogonal frequency division multiple access-based MEC systems in which the time and energy are weightily combined into a system utility that will be maximized through formulating an optimization problem. Then, on the basis of the modified Newton method and the computation offloading priority concept, a low-complexity ranking-based algorithm is designed to obtain the near-optimal solution in an efficient manner. The simulation results proved that the proposed algorithm can obtain near-optimal performance. However, the authors in [149] did not consider the computing resources and the radio resource allocation together which may lead to a significant waste of the other resource [143] (Table 2).

3.2 Offloading Nature

After considering the objectives that effects the offloading decision, the offloading nature can specify which part of the applications should be offloaded for remote execution on the edge server. According to the previous studies, the common models can be classified into the full application or partial application offloading. Full application offloading denotes that the IoT device will offload the entire application to remote server nodes. While application partitioning indicates that only the intensive components are separated and offloaded to remote edge server node for running at run-time.

Table 2 Models objective comparison

Ref.	Objective	Proposed solution
Elgendy et al. [38]	Minimize Time and Energy Consumption	Multiuser Resource Allocation and Computation Offloading Model with Data Security for MEC
Kao et al. [62]	Reduce Latency	Latency Optimal Task Assignment Scheme for Resource-constrained Devices
Khalili et al. [63]	Minimized Weighted Sum of Energy and Latency	Message-Passing Approach for MCC
Liu et al. [80]	Minimize Energy, Execution Delay and Price Cost	Queuing Theory formulation in MEC System
Mao et al. [90]	Minimize Energy Consumption and Execution Delay	Lyapunov optimization-based online algorithm
Ren et al. [108]	Reduce Latency	Three Different Video Compression Offloading Models for MEC
Sheng et al. [118]	Minimize Energy	Energy-Efficient Cooperative Computing for MEC
Xu et al. [137]	Minimize Energy Consumption, Execution Delay and Load Balancing Rate	Multi-objective Computation Model for Internet of Vehicle in Edge-Cloud Computing
Zhang et al. [146]	Minimize Energy	Energy-Efficient Computation Offloading Algorithm
Zhang et al. [149]	Minimize Task Latency and Device Energy Consumption	A Low-Complexity Ranking-Based Algorithm

3.2.1 Full Offloading

This subsection highlights the common models that deal with the application tasks as single components which will be executed locally on the IoT device or will be offloaded and executed remotely at edge server.

A Markov decision process has been utilized to address the task scheduling for MEC systems in [79], where the queueing state of the task buffer queueing state, the transmission unit state and the execution state of the local processing unit are investigated to determine the computation tasks scheduling. In addition, the latency and power consumption of computation task are formulated as an optimization problem with the aim of minimizing the delay. Afterward, an efficient one-dimensional search algorithm is developed to solve this problem and derive the optimal task scheduling policy. Finally, the simulation results proved that the proposed algorithm can achieve a shorter execution delay compared to the other benchmarks. However, protecting the transferred data of mobile application from cyber-attacks is considered the main drawback of this work.

Elsewhere [26, 146], proposed an energy-efficient computation offloading model for multi-user MEC system. Specifically, Zhang et al. [146] proposed a new energy-efficient mechanism for MEC in 5G heterogeneous networks. Besides, they formulate the cost of energy cost for the computing task and file transmission as an optimization problem whose aim is to minimize the system consumption of energy under the constraint of latency. Then, a three-stage offloading scheme is designed to solve this problem in polynomial time and obtain the offloading decision. The numerical results proved the proposed offloading scheme can save the energy consumption by up to 15% in comparison with other baselines. On their part, Chen et al. [26] studied the offloading problem MEC in a multi-channel wireless interference environment. Then, they adopted the offloading decision as a game-theoretic method in which the Nash equilibrium can be achieved using a distributed offloading algorithm. Numerical results demonstrated that the proposed algorithm not only improves the offloading performance (i.e., approximately up to 40% with respect to local execution) but also scales well for large-scale of mobile users.

Recently, Alam et al. [12] proposed an autonomic offloading framework for MEC system, where the resource allocation problem is modeled as a Markov decision process and minimizing the latency of service computing is the main goal. Then, a deep Q-learning-based algorithm is designed to find the optimal offloading decision. The experimental results show that the proposed algorithm can boost the system performance with respect to execution time and energy consumption.

3.2.2 Partial Offloading

This subsection highlights the common models that partially divide the computation tasks on application into two main parts, some of them will be executed locally on the IoT device while the other part will be offloaded and executed remotely at edge server.

In [85], computation offloading and components scheduling are jointly optimized through proposed approach for multi-component mobile applications. Then, a linear optimization problem is formulated with a net utility function that can make a trades-off between energy saved under the constraints of communication delay, the ordering of component precedence and the execution time application. Afterwards, the authors use a real-world data measurements to solve this problem. Finally, the results demonstrated that the proposed approach can decrease the overall overhead by up to 54% and 37% compared to local execution and other existing approaches, respectively. Meanwhile, in [135], computational speed, offloading ratio, and transmit power are jointly optimized through a non-convex problem formulation with the aim of minimizing the energy consumption and application execution latency for IoT devices. Then, a locally optimal search-based algorithm is developed to derive the solution for this problem in an efficient manner. Simulation results verified that the proposed algorithm can significantly minimize the energy consumption and the latency by up to 56% and 23%, respectively, compared with local execution approach. However, a common drawback of [85, 135] is that the data of IoT applications are susceptible to various types of attacks during the transmission process to the server.

Furthermore, other efforts have also considered minimizing the task execution latency [111, 129]. In [111], an offloading scenario in Device-to-Device (D2D) for MEC is considered in which the interference management and the partial offloading are integrated using orthogonal frequency-division multiple access method. Then, energy consumption, resource allocation and partial offloading are jointly considered through a mixed integer nonlinear programming problem formulation. Afterwards, an efficient and novel algorithm is developed to solve this problem in an efficient way. The evaluation results verified that the proposed solution can reduce the execution latency by roughly 70, 42, and 46% with respect to local execution, random offloading and full offloading solutions, respectively. Whereas in [129], Wang et al. formulated the partial offloading application for multi-vehicle system as an optimization problem. Then, an asynchronous advantage actor-critic-based algorithm is designed to solve this problem and obtain the optimal action. Simulation results proved that the proposed algorithm reduces the execution latency by roughly 33%, and 12% with respect to local execution and deep-Q-network solutions, respectively.

3.3 No. of Users

In our taxonomy, depending on the MEC environments, the computation offloading models can be classified into three main categories which are the number of users or IoT devices, the number of edge server nodes and the number of tiers. This subsection addresses the common models regarding the number of users or IoT devices which is classified into two sub-class called single user and multiple user. Whereas, the other two categories (i.e., number of edge server nodes and number of tiers) will be discussed in detail in the following subsections.

3.3.1 Single User

For a single-user computation offloading, Mao et al. proposed a computation offloading approach for MEC system in [88, 89]. More specifically, in [88], a green MEC system has been investigated with energy harvesting devices in which the latency of execution and the failure of the task are jointly optimized through problem formulation. Then, Lyapunov optimization-based algorithm is developed to solve this problem and derive the decision of offloading, the power of transmission and the frequencies of CPU-cycle without demanding information of task request, energy harvesting process and wireless channel. The simulation results verified that the proposed algorithm can reduce the execution time by up to 64% by deploying the computation offloading strategy reasonably. Whereas in [89], the task scheduling and the allocation of transmission power are jointly optimized through a convex optimization problem formulation whose objective is to minimize the weighted sum of the execution delay and energy consumption. The simulation results proved that the task scheduling can achieve a less delay compared with traditional approaches.

Furthermore, a fine-granularity-based offloading strategy is presented in [30] in which minimization of the energy consumption with delay constraint satisfaction is the main goal. In addition, the authors proposed a binary particle swarm optimizers algorithm to solve the problem and derive the task offloading decision. The simulation results show that the proposed algorithm can save the energy consumption of IoT device by up to 25% with respect to the local execution approach.

Recently, an efficient offloading scheme is proposed in [96] where the cloud and edge computing resources are jointly cooperated for IoT. In addition, IoT devices competitive on the computation resources are formulated as an integer problem whose objective is to reduce the execution latency. Then, an iterative and heuristic algorithm is developed to derive the solution for this problem in an efficient manner. Finally, the simulation results show that the proposed algorithm can decrease the execution latency by at most 30% with respect to the other baseline algorithms. However, the energy consumption constraint is not considered in this study.

3.3.2 Multiple User

Regarding a multi-user computation offloading, in [74] an energy-aware offloading approach is introduced for IoT devices over heterogeneous networks, in which the computation resources, power consumption, states of channel and delay requirements are jointly optimized through problem formulation. Afterwards, the original problem is decomposed into two sub-problems and then an iterative framework is proposed to derive the power allocation and offloading decision efficiently. The simulation results verified that the proposed approach can find a near-optimal solution. In [33], a multi-user computation offloading problem is formulated as graph cut problem and spectral clustering-based solution is proposed to solve this problem. Specifically, the network graph of the mobile applications is firstly defined and the label propagation theory is investigated to simplify this graph and derive the optimal solution. The experimental results demonstrate that the proposed approach can minimize the energy and transmission consumption during the offloading process.

Minimizing energy consumption and delay constraint have been considered in [40, 58]. In [58], the task offloading and the allocation of bandwidth are jointly formulated as an optimization problem for multi-user MEC system in which minimizing the overall cost of delay and energy is the main goal. Then, a Deep-Q-Network-based algorithm is developed to solve this problem and obtain the close-to-optimal solution. The numerical results proved that the proposed algorithm can reduce the total cost by about 25% with respect to local execution scenario. Meanwhile, Elgendy et al. [40] presented an efficient and secured offloading model for multi-user multi-task MEC system in which resource allocation, task offloading, security and compression issues are jointly formulated as optimization problem which aims to minimize the weighted sum of energy under a latency constraint, shown in Fig. 7. Besides, the

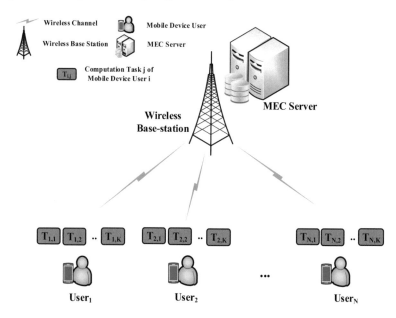

Fig. 7 System model of [40]

authors utilized JPEG and MPEG4 as a compression layer to decrease the overhead of the transferred data. Afterward, an efficient algorithm is designed to derive the offloading decision through utilizing linearization and relaxation approaches. The simulation results verified that the proposed model can save the total cost by about 46% with respect to local execution scenario. In addition, it can scale well for the large number of IoT devices.

3.4 No. of Edge Nodes

Regarding number of edges server nodes, the computation offloading models are classified into two main categories which are single node and multiple nodes. In a single node environment, the computation tasks of IoT application can be offloaded to only a single edge server node. Whereas in a multiple nodes environment, there are a set of edge server nodes in which the application's tasks can be offloaded and executed. In the reminder of this section, the common models are highlighted according to these two categories.

3.4.1 Single Node

For a single node environment, Zhao et al. [150] maximized the number of IoT applications that are served by edge computing nodes while the delay requirements for
these applications are satisfied. Specifically, the authors proposed a new cooperative
scheduling model in which the application data and code are sent to the scheduler
(i.e., main component of this model), shown in Fig. 8. Then, the scheduler will decide
whatever the offloaded application will be executed on the virtual machine (VM) of
MEC or it will be delegated to the distant Internet Cloud. This decision is depending
on two constraints, the computing resources's availability at edge computing node
and the applications priorities (where the low delay's application has a higher priority). Regarding the former constraint, if there is an empty virtual machine node with
sufficient resources, the application will be allocated and processed at this node and
finally the result is sent back to the UE. For the second constraint, where, the power
of computation introduced via edge computing server is not suitable, the scheduler
delegates the application to the distant Internet Cloud. Further, a priority-based cooperation policy is presented, in various thresholds of buffer are defined for each priority
level. Therefore, in the full buffer case, the applications are sent to the internet cloud.
It also uses a low-complexity recursive algorithm for determining the local optimal
threshold. The authors compared his work with three other policies which are MEC
only, First Come First Serve-based Cooperation strategy and non-buffer strategy.
The proposed strategy can boost the application completion probability within the
tolerated delay by 25%. However, treating the mobile application as a single unit
is considered one of the main disadvantages of this model, where there is another
situation (such as codes that need mobile hardware like GPS or camera) which is
better to divide the application into tasks and to offload only the intensive tasks.

Fig. 8 System model of
[150]

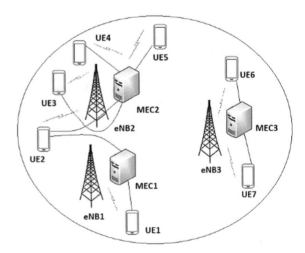

Moreover, maximizing the utilization of the VMs on MEC servers which taking eNBs overload, network delay and VMs migration cost as constraints is the main goal of [31], in which the authors formulated the VM allocation problem as a markov decision process and then proposed an algorithm to compute the optimal policy for allocating the suitable VM to each IoT device. Finally, the results proved that when the VM migration cost increases, it's better to be allocated locally at the serving eNB. Meanwhile, in [52], reducing the execution delay for the offloaded applications and minimizing the total power consumption of MEC are considered the main objectives, in which the authors considered a hot spot area densely populated by the IoT devices, which can access various MEC servers via the closer eNBs. They also assumed that each MEC server is connected to only one of the nearby eNBs. Regarding the simulation results, the proposed index policy is more costly than optimal policy by 6.5% in the worst case in terms of system cost. Nevertheless, in both [31] and [52], the authors did not consider more computing nodes for each application to further decreasing the execution delay. In addition, in [31], the energy consumption does not consider in their model. Whereas, in [52], the interference constraint is not considered.

Elsewhere, [49, 151] introduced offloading solutions for MEC systems in which, minimizing the total energy consumption subject to the delay requirements of tasks is the main goal. Specifically [151], proposed a computation offloading model for a multi-user MEC system and then, developed an algorithm which can allocate the computational resources and the system bandwidth for IoT devices. Further, this algorithm is efficiently approximated with energy discretization to reduce its complexity and got the near-optimal solution. The simulation results demonstrated that the proposed model can save the energy consumption by approximately 82.7% with respect to local execution approach. Meanwhile, in [49], the communication and computational resources are jointly optimized through a non-convex problem formulation. Then, this problem is decomposed into two sub-problems to derive the optimal solution efficiently. However [49, 151], had their own drawbacks, in which both of them suffering from securing the transferred data from different types of attacks.

3.4.2 Multiple Nodes

With regard to multiple nodes environment, Oueis et al. [99] addressed the effective of eNBs cooperation on the cluster's characteristics in terms of the execution latency and the power consumption constraints where three different optimization clustering strategies are proposed. The authors used a single floor building with 10m x 10m apartments in a 5 × 5 grid to evaluate these strategies, where each apartment is assumed to be equipped with an eNB, shown in Fig. 9. Then, these eNBs is classified into three sets which are serving eNB located at the grid center, eNBs that are separated from the serving eNB by at most two walls and finally the remaining are far eNBs. In the first strategy, it forming a cluster of eNBs in order to reduce the overall latency. Therefore all of the active eNBs nodes are forced to Participate

Fig. 9 System model of [99]

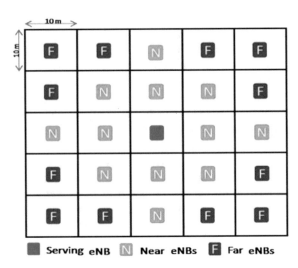

■ Serving eNB N Near eNBs F Far eNBs

in the computation cluster. However, this strategy achieves up to 22% reduction of cluster latency, but unfortunately, it does not consider any power consumption as constraints. Therefore, the second strategy handled this problem by taking the overall power consumption of cluster as constraints. In this case, serving eNBs nodes preferred in computation to minimize the overall power consumption of the eNBs. However, this strategy achieves up to 61% reduction of power consumption, but unfortunately, it may be lead to increase the energy consumption for some eNBs nodes than others. As a result, the last strategy overcomes this problem by distributing the power consumption on eNBs in equality and if there is eNB that has a greater power consumption than the others, the load distribution can be modified, if possible.

Compared to the previous paper, Oueis et al. extended their work by addressing a multi-user case in [100], in which minimizing the cluster power consumption and guaranteeing the required execution delay for each IoT device is the main objective. The authors used the same deployment model and formulated an optimization method where the cluster of eNBs is more scalable depending on the IoT device application request and its requirement. The main idea of this study is to jointly compute the clusters for all active IoT devices' requests simultaneously to be able to efficiently distribute the computation and communication resources among the IoT devices as well as to achieve higher quality of experience. The results of this model is compared with other three different scenarios which are no clustering, static clustering and successive clusters optimization. Regarding the Users satisfaction ratio, the proposed model can achieve up to 94% of IoT devices to be satisfied which is better than the other scenarios. But the average power consumption per IoT device is significantly higher than no clustering and successive clusters optimization scenarios. Also, the average latency gain is better in successive clusters optimization scenarios than the proposed one.

Fig. 10 System model of
[122]

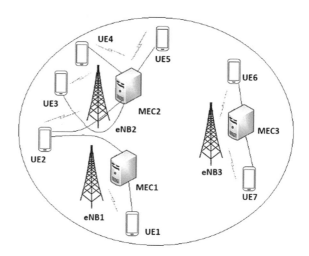

Moreover, Tanzil et al. [122] proposed a new model that maximally exploits MEC local computational resources while minimizing the remote cloud utilization, shown in Fig. 10. This model forms a cluster of eNBs neighbors' resources based on a canonical cooperative game approach while these eNBs can share their resources with each other. Each eNB node can take monetary incentives if they process the computation for the IoT devices that are attached to other eNB in this cluster. This coalitions is formed for several time slots and then new coalitions may be created. The authors formulated optimization problems with a utility function to deal with the eNBs resource sharing problem. This model is compared with isolated eNBs (where there is no cooperation between eNBs and each node operates individually) and Grand femto-cloud (where all the eNBs in the system are participating in the computation). The results show that the proposed approach is better than the other scenarios in execution delay reduction. A common drawback in [122] is that the formation of new coalitions is considered in the proposed scheme.

Furthermore, an online application placement scheme for a MEC environments is proposed in [132] in which, the IoT device's application and their physical computing system are modeled as graphs and the computation resources (i.e., availabilities and demands) are annotated. Then, an exact algorithm is designed for linearly assigning the application graph onto the tree of the physical graph in which the node and link assignment as well as incorporating multiple types of computational resources at nodes are jointly considered. Afterwards, this algorithm is efficiently approximated with a polynomial-logarithmic competitive ratio for the placement of tree application graph. The theoretical evaluation proved that the proposed algorithm can achieve practically well on average. However, the application data transferred is not sufficiently safeguarded from different types of attacks during transmission.

More recently, Computation-Communication Tradeoff model for multiple edge computing networks is presented in [72], where the computation tasks are replicated onto multiple edge nodes and then multiple copies of the results are copied which can speed up the downloading phase through the transmission cooperation. The numerical results demonstrated that the downloading time decreases proportionally and linearly in the binary and partial offloading cases, respectively. However, this study did not address the load balancing issue among multiple edge nodes which has a great impact on reducing the communication delay end energy consumption of IoT device. Moreover, the data of application is not protected from different attacks during the transmission process.

3.5   No. of Tiers

Depending on the number of layers in MEC environments, the computation offloading models can be classified into two main categories which are single tier and multi-tiers environment. In the following subsections, the common models regarding the number of tiers will be discussed in detail.

3.5.1   Single-Tier

Considering a single-tier computation offloading models, Bi et al. [22] proposed an offloading model for multi-user MEC system in which the whole computation tasks is treated as single unit that will be executed locally or remotely. This model jointly optimized the mode selection of individual computing and the transmission time allocation with the goal of maximizing the rate of computation for all IoT devices. Finally, the extensive simulation results proved that the proposed model can significantly outperform the local execution and full offloading scenarios by nearly 18.5% and 26.2% respectively. However, considering the computation tasks as a single unit to be offloaded or not is the main drawback of this work, in which huge amounts of data will be transmitted over the network. Moreover, the application data is not protected during the transmission process.

Deep reinforcement learning-based algorithms are utilized in [57] and [56] to address the computation offloading model for a multi-user MEC system. In [57], the allocation of bandwidth and offloading decision are jointly optimized through an integer problem formulation whose objective is to minimize the total delay for completing tasks and the corresponding energy. Then, a distributed deep learning-based algorithm is designed to tackle the curse of dimensionality for this problem and to derive the offloading decision in an efficient manner. The numerical results verified that the proposed algorithm can save by up to 39.4 and 21% of total cost with respect to the local and edge processing scenarios. Similarly, in [56] Huang et al. formulated an integrated model for allocating the wireless resources and the task offloading whose objective is to maximize the weighted sum of computation rate. Afterwards, an online deep leering-based algorithm is designed to obtain the near-optimal offloading decision efficiently. Nonetheless, a common drawback in both approaches above (i.e., [56, 57]) is that applications' data of IoT devices are vulnerable to different types of attacks during the transmission process.

Furthermore, offline and online optimization solutions are proposed for a single user with one multi-antenna MEC system [127], in which the task offloading and energy allocation are jointly with the objective of minimizing the transmission energy consumption. First, a priori knowledge of task and channel state information are utilized to obtain the optimal solution. Then, this optimal solution is utilized to further propose a heuristic and online approach which can derive a near-optimal solution for static and time-varying channels scenarios. Finally, the numerical results proved that the proposed solution can significantly reduce the energy consumption with respect to the other benchmark solutions.

3.5.2 Multi-tier

Considering a multi-tiers computation offloading models, Meng-His et al. [24] jointly optimized the computation and communication resources and the decision of offloading for a multi-user MEC system through an integer problem formulation whose objective is to minimize the total system cost in terms of energy, computation, and delay. Then, semi-definite relaxation, alternating optimization and sequential tuning are utilized to solve this problem and derive the local optimum solution through a three-step algorithm in an efficient way. The numerical results proved that the proposed model can save the overall cost by up to 29.6% with respect to the local execution approach. However, the transmitted data of mobile applications is endangered during the transmission process.

Elsewhere [74, 77], introduced task offloading solutions for mobile edge cloud computing systems to effectively minimize the total energy consumption of IoT devices. Specifically [77], proposed a cloud-assisted edge computing model and formulated an integer optimization problem in which minimizing the energy consumption subject to the completion time deadline and task-dependency requirement is the main goal. Then, a semi-definite relaxation and stochastic mapping-based algorithm is proposed to solve this problem and obtain the offloading decision efficiently. The simulation results proved that the proposed algorithm can reduce the energy consumption by approximately 38.5 and 11.1% with respect to local and cloud processing scenarios. Meanwhile, Li et al. [74] formulated an energy-aware offloading model for a multi-user environment where the heterogeneity of the computation resources, the requirement of latency, the power consumption at IoT devices and the states of channel are jointly considered. Afterwards, an iterative solution is introduced to solve this problem and allocate the transmission power and the task offloading policies. The simulation results verified that the proposed framework is competitive with respect to the optimal solution. Nonetheless, balancing the load between edge server nodes is not addressed in these studies which has a great impact on reducing the communication delay end energy consumption of IoT devices.

Recently, in [59, 148] multi-server, multi-user, multi-task computation offloading models are presented for edge-cloud computing systems. In [59], Huang et al.

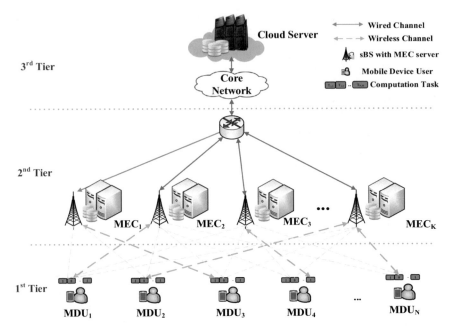

Fig. 11 System model of [148]

formulated different real-time task policies as an optimization problems and relaxation and deep learning-based algorithms[14, 54, 117] are proposed to derive the close-optimal offloading decision efficiently. Then, an extended algorithm is developed to improve the convergence performance. The numerical results proved that the proposed algorithm can efficiently derive the close-optimal solution in less than 1 millisecond. Further, Zhang et al. [148] jointly considered the load balancing, security and task offloading for edge-cloud computing system, shown in Fig. 11. First, an efficient load balancing algorithm is developed to balance the load between edge server nodes based on the number of IoT devices, data rate and central processing unit's cycles for each task. In addition, a novel layer of security is introduced in which advanced encryption standard with electrocardiogram signal-based key technique is utilized to protect the application data from different attacks. The simulation results verified that the proposed scheme can save the system cost by approximately 68.2% with respect to local execution. However, the main disadvantage of [59] is that the applications' data are not protected from different attacks during the transmission process.

3.6 *Mobility Concern*

The mobility concern category of our taxonomy deals with the management of the computation tasks offloading through the movement of IoT devices between different edge nodes. The common offloading models used three different approaches to address the mobility which are power control, virtual machine migration and path selection. In the power control approach, the transmission power of the edge server node is adapted during the execution of the offloaded tasks and this approach is applicable for the limited mobility of IoT devices (e.g., IoT devices are slowly moving inside a building). Whereas the virtual machine migration approach migrates the virtual machine which is responsible for executing the offloaded tasks to another, more suitable, computing node and this approach is applicable when the IoT devices mobility is not limited (e.g., IoT device performs handover to the new serving node). Finally, in the path selection approach, the offloading model is responsible for selecting a new communication path between the IoT device and the computing node which is more suitable and this approach is applicable when a huge amount of data needs to be migrated among the computing nodes. In the following subsections, the common offloading models regarding the mobility concern will be discussed in detail.

3.6.1   Power Control

Considering the power control approach, Mach et al. [82] proposed an efficient cloud-aware algorithm for adapting the transmission power of small cells with the objective of maximizing the number of processed applications at MEC under latency constraint. In particular, the small cell can adapt (i.e., increase or decrease) the transmission power for the IoT device's movement using coarse and dynamic fine settings to avoid the handover to a new cell before the result of the offloaded task is received, if possible. The simulation results proved that the proposed algorithm can successfully increase the number of deliverable applications to IoT device with up to 95% in contrast with non-cloud-aware which achieves approximately 80%. An extension of [82] is given in [83] in which the time for the transmission power adjustment is applied individually for each IoT device with respect to the channel quality which increased the ratio of deliverable applications to reach 98%. However, the main drawback of [82] is the late adaptation of the power control when sufficient channel quality is not guaranteed in the required time.

Furthermore, the transmission power control and virtual machine migration are jointly utilized to manage the IoT device mobility and maximize the scalability for a multi-user multi-cloudlet systems [109]. More specifically, the processing delay, backhaul delay, and the transmission delay are formulated as a mathematical model for MEC, then an efficient and heuristic algorithm based on particle swarm optimization technique is developed to balance the workload among the cloudlets and then boost the efficiency of MEC. The experimental results demonstrate that the proposed scheme not only outperforms the traditional methods in the number of serviced users but also can deliver a scalable service with the different types of requirements scenarios.

3.6.2 Virtual Machine Migration

Regarding the VM migration approach, the decision process of VM migration for a multi-user with multi-service instance mobile micro-cloud computing system has been studied [131]. The authors proposed an offline algorithm to find the best configuration for the placement based on a-priori knowledge of the instance arrivals and departures within a look-ahead window. Afterwards, an online algorithm is proposed to solve this problem in real-time and with polynomial time-complexity. Finally, an efficient method is proposed to optimally find the look-ahead window size with the goal of minimizing the upper bound of the placement cost. Simulation results verified that the proposed offline and online algorithms can minimize the cost by up to 25%, 32% and 32%, 50% with respect to non-migration and always migration policies, respectively. It addition, they are applicable for a larger class of dynamic resource allocation problems. Moreover, maximizing the system throughput of distributed data-centers is considered in [116], in which Locator/Identifier Separation-based architecture protocol is proposed for cloud access improvement. The VM migration process to a new edge node will be determined on the basis of the required and the available computing resources at edge server nodes as well as with a given latency or jitter threshold. Finally, the simulation and real evaluations verified that the proposed solution can significantly improve the throughput of the system by roughly 40% with respect to non-migration policy. However, these studies [116, 131], do not consider the optimization of consumed energy of IoT devices during task execution which is considered as the main metric in the decision offloading process. In addition, data of mobile applications are not protected from attacks during the transmission process.

Elsewhere [93], presented an effective solution to improve the migration process through proposing a mobility-based prediction approach. This approach handled the trade-off between quality of experience and total overhead, in which the IoT service request is spitted into several portions that will be optimally processed at the most suitable micro data centers based on estimating the throughput of data transferring between datacenter and IoT device as well as estimating the time windows for performing IoT device handover. The simulation results verified that the proposed approach can reduce the transfer latency by roughly 35% compared with recent related approaches. In addition, Sun et al. [121] proposed a cloudlet network architecture for mobile cloud computing to reduce the E2E delay and meet the quality of service requirements. Then, migration cost and gain have been jointly optimized through a mixed-lingerer problem formulation whose objective is to maximize the total profit of the live migration process. Afterwards, a heuristic solution is derived using a mixed-lingerer quadratic programing-based tool. Finally, the simulation results demonstrate that the proposed architecture can reduce the execution delay and migration cost by approximately 90% and 40%, respectively, with respect to non-migration approach. However, these studies (i.e., [93, 121]) do not address the load balancing issue between edge computing nodes. Moreover, like in [116, 131], IoT applications' data are not adequately protected during the transmission process.

Recently, Ali et al. [15] proposed a novel energy-efficient resource allocation approach for MEC system. This approach is optimally assigned the task request of IoT device to the most suitable server and allocated the proper resources for IoT device. In addition, a low-cost energy-efficient algorithm based on deep learning is designed to handle the complexity of multidimensional resource allocation problem, in which the different load of IoT devices' request and their heterogeneous nature are jointly considered. The simulation results demonstrate that the proposed algorithm can significantly minimize the energy consumption and boosts the service rate by up to 26 and 23% with respect to other algorithms.

3.6.3 Path Selection

With regard to the path selection approach, Zdenek et al. [21] proposed a path selection algorithm for transferring the data of the offloaded task between small cell and IoT device with the goal of minimizing the energy consumption and transmission latency. Further, the handover process is enforced to the new serving cell to achieve this goal. Finally, the simulation results verified that the proposed algorithm can minimize the transmission delay by up to 9% in comparison with the traditional delivery schemes. In addition, the satisfaction of IoT device with experienced delay is increased by approximately 6.5%. An extension of [21] is given in [103] in which the complexity of the proposed algorithm is decreased, thereby, the transmission delay is minimized by up to 54.3% while the satisfaction of IoT device with experienced delay is increased by approximately 28%. A common drawback in both approaches above (i.e., [21, 103]) is that the proposed algorithm is not sufficient for the case of the distance between the computing location and IoT device is too far.

Similar to the enumerated efforts, Plachy et al. [104] proposed a markov decision process-based algorithm for managing the IoT device's mobility in MEC systems. Specifically, the computation and communication resources' load of edge node as well as the movement prediction of the IoT device are dynamically exploited to determine the VM placement and to derive the most suitable path between the IoT device and edge node for returning the results back to IoT device. Simulation results demonstrate that the proposed algorithm can reduce the offloading time by roughly 10% compared to state of the art approaches. Further, Yu et al. [142] proposed a dynamic mobility-aware partial offloading algorithm for MEC system. In this algorithm, the offloaded data are assigned to the most suitable edge node for processing based on the prediction of IoT device's movement in which minimizing the energy consumption with delay requirements satisfaction is the main goal. The simulation results show that the proposed algorithm can significantly save about 70% of energy consumption with respect to the traditional approaches.

3.7 Security Concern

Regarding the security concern item of our taxonomy, the computation offloading models can be classified into three main categories which are authentication, encryption and privacy. In the following subsections, the common models regarding the security concern item will be discussed in detail.

3.7.1 Authentication

Considering the authentication policy, Ibrahim et al. [60] proposed an efficient and secured mutual authentication approach for edge-cloud computing system, in which the IoT device's user can be authenticated [55] with any server node at the fog layer using only one long-lived master secret key. In addition, the IoT device's user does not need to re-register or re-initialize with a new fog node, which reduces the extra overhead, thereby is considered as a more suitable solution to be applied on the Fog user's smart device. Meanwhile, in [19], a privacy-preserving authentication solution is proposed for edge-fog computing system. They used pseudonym-based cryptography technique to preserve the IP while single fog node's authentication, intra-fog and inter-fog authentication are utilized to support the high mobility of IoT device. In addition, bilinear pairing is used to generate the session key which can offer the data confidentiality. However, a common drawback in both approaches above (i.e., [19, 60]) is that the departure from the old nodes is not considered; thereby, the mobility support is still limited. In addition, transferring the IoT device's identity as plaintext is considered another drawback of [60], which makes the communication prone to different attacks and users' privacy is revealed.

A novel lightweight mutual authentication solution is proposed in [128] for N-times computation offloading in IoT, in which the IoT device and edge node are authenticated anonymously. Subsequently, the privacy of the device's user is preserved. In addition, lightweight one-way hash and MAC function are utilized for signing/verifying process and for protecting the user from denial of service attack. Finally, the performance analysis verified that the proposed solution can respectively boost the computing speed of IoT and edge devices by 1.66x and 2.87x.

Recently, a lightweight authentication protocol for edge-based applications is proposed in [94], in which, the edge entity derived the idea of hash function to broadcast the authenticity, integrity, confidentially, and forward secrecy messages using session keys. In addition, the authenticated encryption and hash chains are utilized to make this protocol resilient for quantum attacks. Finally, the security analysis and comparative evaluations proved that the proposed protocol require only 96 and 56 bytes for storage and communication overhead, respectively.

3.7.2   Encryption

With regard to the encryption approach, Tawalbeh et al. [124] proposed an efficient and secured framework for mobile cloud computing, in which the data of IoT devices are encrypted on the basis of their confidentiality. Specifically, the data confidentiality are classified into three levels which are basic, confidential or highly confidential. Then, Hypertext Transfer Protocol Secure (HTTPS) and Transport Layer Security (TLS), AES-128 or AES-256 are respectively utilized to encrypt the transmitted data. The simulation results proved that the proposed framework can ensure the data integrity and confidentiality. However, classifying the data adding extra overhead on the IoT devices which is considered a drawback of this framework.

Elsewhere [40, 65], introduced an efficient and secured solutions for computation offloading of edge-cloud computing systems. More specifically [40], introduced a new layer of AES with a genetic algorithm to protect the transferred data from different attacks In addition, MPEG4 and JPEG techniques are utilized to compress the transmitted data. Thereby, reducing the communication overhead. Furthermore, an offloading algorithm is developed to find the cost-efficient solution optimally. Finally, the experimental results proved that the proposed approach can reduce the total overhead of the system by roughly 46% with respect to local execution. Meanwhile, in [65], an efficient and secured approach is presented for unmanned aerial vehicle-enabled edge-cloud computing system, in which the data of each task are protected from cyber-attacks through utilizing chaotic substitution Box with highly dispersive. In addition, the authors formulated the offloading and resource allocating as an optimization problem in which minimizing the consumption of energy is the main goal. Furthermore, two heuristic algorithms are developed to derive the sub-optimal solutions for this problem. The experimental results verified that the proposed approach can reduce the consumption of energy by up to 14% with respect to other benchmark solutions.

More recently, Zhang et al. [148] proposed an optimized and secured approach for a multi-server edge-cloud computing system, in which AES technique boosted ECG signal is introduced to protect the application data from cyber-attacks. In addition, a new load-balancing algorithm is developed to redistribute the load between edge nodes. The simulation results verified that the proposed approach can reduce the overall consumption by roughly 68.2% with respect to local execution.

3.7.3   Privacy

Considering the data privacy for edge-cloud computing, Xu et al. [138] proposed an offloading approach for internet of vehicles, in which the privacy of application data is preserved. Specifically, vehicle-to-vehicle routing method is designed which based on the communication between vehicles. Afterwards, tasks' execution time, edge's energy and privacy satisfaction objectives are achieved though non-dominated sorting genetic algorithm utilization. The simulation results validated that the proposed approach is more effective for edge devices. However, the data of application are not protected from cyber-attacks.

Other efforts have also preserved the privacy through the offloading process [86, 153]. In [153], a hidden markov model is utilized to preserve the privacy of the time-series classification activities of the patients' data in e-healthcare systems. In addition, a new protocol is introduced and an algorithm is developed for preserving the result of classification. The performance evaluations proved that the proposed approach can preserve the privacy for the classification activities as well as decrease the cost of computation. On their part, Zheng et al. [86] introduced an alternative deep neural networks-based training strategy for resource-limited users. In addition, a differential privacy-based algorithm is developed to ensure the level of activation and to provide the privacy for the user. Further, a federated learning-based approach is presented to parallel the training process which concerning the privacy. The simulation results verified the possibility of offloading solutions for deep neural network models regarding solo and parallel modes. Nevertheless, it's not practical to assume that the central server has infinite resources of computing as discussed in [86].

4 Challenges and Open Issues

As seen in the preceding sections, the computation offloading in edge-cloud computing systems has gained a lot of attention because of its ability to alleviate the limitations of IoT devices. Despite recent developments and significant efforts in the field of computation offloading in edge-cloud computing, there are many important challenges and open issues exist which need to be handled. In this section, we mention and discuss several challenges and open issues that need to be handled and can guide the future research in this field.

4.1 Computation Offloading for a Large-Scale Networks

The computation offloading for a large-scale edge-cloud computing systems has been addressed in some research papers using traditional and machine learning methods [27, 70]. However, there are still some challenges that have not been addressed. As we know when the number of IoT devices and edge computing nodes increases, the complexity for the offloading models increases and then reflect on the offloading decision delay. Specifically, the process of offloading decision is centralized in which all the information about IoT devices, their computation tasks, channel bandwidth and edge nodes resources are collected and then the offloading decision is determined. Therefore, for large-scale network, the network channel will be overloaded due to the gathering information process in which the current works neglect the time consumed by this process. In addition, the time-complexity for solving the offloading models increased with the network scale. Over and above, the real-time systems with the dynamic and large-scale networks (e.g., e-health infrastructures and intelligent transportation systems) introduce a new challenge, in which the information will be captured from heterogeneous devices.

4.2 Mobility Management for Computing Offloading

As mentioned in Sect. 3.6, the management for mobility issue in the computation offloading models has been studied in many papers, in which mobility of the users are modeled using predication or probability methods. However, most of the previous studies neglect to address the computational resources' effects at edge computing nodes due to different handover strategies. In addition, managing the mobility in heterogeneous architecture is not addressed in which IoT device users can make frequent handovers between different architectures of edge nodes (e.g., WiFi access points, macro-cell and small-cell base stations), thereby can lead to extremely complicated model because of the diversity of system configurations. Moreover, the channel interference will be occurred due to the movement between different cells which will significantly decrease the communication performance as well as deteriorate the quality of experience. Finally, balancing the load between different cells due to the mobility of IoT devices has not been adequately addressed.

4.3 Software Defined Network and Edge-Cloud Computing Integration for Computation Offloading

Most of the current studies assume that the edge server nodes have unlimited computational resources which are unpractical assumption. Moreover, edge server nodes need to be cooperated and worked together to offer satisfactory service. However, the interoperability or inter-cooperation between edge nodes poses a new challenge of management. Software Defined Network is considered as an efficient solution to manage the inter-cooperation between edge server nodes and can abstract the internal complexities of the interoperability from the users [106]. Recently, software defined network and edge-cloud integration have gained a lot of attention, in which some studies have introduced this integration and discussed their effect on the services' improvement [17, 18]. However, Software Defined Network and edge-cloud integration for computation offloading have some challenges and open issues that need to be addressed, particularly for modeling the computation offloading. In addition, distributed and centralized software defined network should be considered in the computation offloading models. Furthermore, the security and quality of service issues need to be addressed.

4.4 Security and Privacy for Computation Offloading

Security and privacy are considered as critical issues for computation offloading, in which the benefits of edge-cloud computing paradigm would be overshadowed without a proper and more effective security and privacy mechanisms due to the impairment caused by different types of threats and cyber-attacks. Even though there are a lot of studies have addressed the security and privacy concerns for computation offloading models, there are challenges and open issues still exist. First, the instinctive heterogeneity of edge-cloud computing systems, due to the different types of edge servers and their vendors, makes the traditional mechanisms are not applicable. In addition, the diversity of communication technologies can bring new types of security threats such as distributing the credentials to all the network elements. Moreover, migrating the private and sensitive information of computation tasks among different edge server nodes, due to the mobility of IoT device, made the data integrity and privacy are hard to guarantee.

4.5 Artificial Intelligence and Deep Learning Integration for Computation Offloading

Recently, Artificial Intelligence techniques and deep learning algorithms have been utilized as effective and more powerful approaches to cope with computation offloading modelling in edge-cloud computing systems, in which they provide accurate decisions and better portability and adaptability [75, 134]. In addition, few studies have used deep learning to address the resources' scheduling and optimization at the edge computing. However, most of the current studies are at the theoretical level in which the deep learning training' process is simulated at the edge node. Moreover, there are still several challenges and open issues that need to be discussed such as exploring the overhead consumption of the system bandwidth due to the data transmission training and thereby the processing latency caused by deep learning inference. Besides, the performance analysis of deploying deep learning algorithms and artificial intelligence techniques in practical edge computing networks for long-term online resource orchestration has not been adequately addressed. Furthermore, the security issues based on deep learning techniques should be more explored, in which blockchain is still in its infancy [1, 152].

5 Conclusion

Edge-cloud computing is considered as a new paradigm for addressing the resource-limitation problems of IoT devices through computation offloading concept in which the computation and storage capabilities of cloud computing have been provided at the edge of the network that enables mobile applications to be processed as well as satisfying the delay requirements. To this end, compared to other surveys, this paper provided a comprehensive survey of state-of-the-art MEC research with a focus on computation offloading on edge-cloud computing combination. In addition, we provided a novel taxonomy on computation offloading at edge-cloud computing combination and introduced the most and common recent computation offloading models regarding this taxonomy. Furthermore, we highlighted the main strengths, weaknesses and other issues which require further consideration. Finally, open research challenges and issues and new research trends in edge-cloud computing are discussed which can guide the future research in this field.

References

1. Abbas, K., Tawalbeh, L.A., Rafiq, A., Muthanna, A., Elgendy, I.A., El-Latif, A., Ahmed, A.: Convergence of blockchain and IoT for secure transportation systems in smart cities. Secur. Commun. Netw. **2021** (2021)
2. Abd El-Latif, A.A., Abd-El-Atty, B., Mazurczyk, W., Fung, C., Venegas-Andraca, S.E.: Secure data encryption based on quantum walks for 5G internet of things scenario. IEEE Trans. Netw. Serv. Manag. **17**(1), 118–131 (2020)
3. Abd El-Latif, A.A., Abd-El-Atty, B., Mehmood, I., Muhammad, K., Venegas-Andraca, S.E., Peng, J.: Quantum-inspired blockchain-based cybersecurity: securing smart edge utilities in IoT-based smart cities. Inf. Process. Manag. **58**(4), 102549 (2021)
4. Abd El-Latif, A.A., Abd-El-Atty, B., Venegas-Andraca, S.E., Elwahsh, H., Piran, M.J., Bashir, A.K., Song, O.Y., Mazurczyk, W.: Providing end-to-end security using quantum walks in IoT networks. IEEE Access (2020)
5. Abd EL-Latif, A.A., Abd-El-Atty, B., Venegas-Andraca, S.E., Mazurczyk, W.: Efficient quantum-based security protocols for information sharing and data protection in 5G networks. Future Gener. Comput. Syst. **100**, 893–906 (2019)
6. Abd El-Latif, A.A., Li, L., Wang, N., Peng, J.L., Shi, Z.F., Niu, X.: A new image encryption scheme for secure digital images based on combination of polynomial chaotic maps. Res. J. Appl. Sci. Eng. Technol. **4**(4), 322–328 (2012)
7. Abolfazli, S., Sanaei, Z., Ahmed, E., Gani, A., Buyya, R.: Cloud-based augmentation for mobile devices: motivation, taxonomies, and open challenges. IEEE Commun. Surv. Tutor. **16**(1), 337–368 (2014)
8. Abou-Nassar, E.M., Iliyasu, A.M., El-Kafrawy, P.M., Song, O.Y., Bashir, A.K., Abd El-Latif, A.A.: Ditrust chain: towards blockchain-based trust models for sustainable healthcare IoT systems. IEEE Access **8**, 111223–111238 (2020)
9. Afrin, M., Jin, J., Rahman, A., Rahman, A., Wan, J., Hossain, E.: Resource allocation and service provisioning in multi-agent cloud robotics: a comprehensive survey. IEEE Commun. Surv. Tutor. (2021)
10. Ai, Y., Peng, M., Zhang, K.: Edge computing technologies for internet of things: a primer. Digit. Commun. Netw. **4**(2), 77–86 (2018)
11. Al-Shuwaili, A., Simeone, O.: Energy-efficient resource allocation for mobile edge computing-based augmented reality applications. IEEE Wirel. Commun. Lett. **6**(3), 398–401 (2017)
12. Alam, M.G.R., Hassan, M.M., Uddin, M.Z., Almogren, A., Fortino, G.: Autonomic computation offloading in mobile edge for IoT applications. Future Gener. Comput. Syst. **90**, 149–157 (2019)

13. Alanezi, A., Abd-El-Atty, B., Kolivand, H., El-Latif, A., Ahmed, A., El-Rahiem, A., Sankar, S., S Khalifa, H., et al.: Securing digital images through simple permutation-substitution mechanism in cloud-based smart city environment. Secur. Commun. Netw. **2021** (2021)
14. Alghamdi, A., Hammad, M., Ugail, H., Abdel-Raheem, A., Muhammad, K., Khalifa, H.S., Abd El-Latif, A.A.: Detection of myocardial infarction based on novel deep transfer learning methods for urban healthcare in smart cities. Multimed. Tools Appl. 1–22 (2020)
15. Ali, Z., Khaf, S., Abbas, Z.H., Abbas, G., Muhammad, F., Kim, S.: A deep learning approach for mobility-aware and energy-efficient resource allocation in MEC. IEEE Access **8**, 179530–179546 (2020)
16. Alshahrani, A., Elgendy, I.A., Muthanna, A., Alghamdi, A.M., Alshamrani, A.: Efficient multi-player computation offloading for VR edge-cloud computing systems. Appl. Sci. **10**(16), 5515 (2020)
17. Amadeo, M., Campolo, C., Ruggeri, G., Molinaro, A., Iera, A.: SDN-managed provisioning of named computing services in edge infrastructures. IEEE Trans. Netw. Serv. Manag. **16**(4), 1464–1478 (2019)
18. Amadeo, M., Campolo, C., Ruggeri, G., Molinaro, A., Iera, A.: Towards software-defined fog computing via named data networking. In: IEEE INFOCOM 2019-IEEE Conference on Computer Communications Workshops (INFOCOM WKSHPS), pp. 133–138. IEEE (2019)
19. Amor, A.B., Abid, M., Meddeb, A.: A privacy-preserving authentication scheme in an edge-fog environment. In: 2017 IEEE/ACS 14th International Conference on Computer Systems and Applications (AICCSA), pp. 1225–1231. IEEE (2017)
20. Baxter, M.: The five pillars of edge computing (2019). https://www.information-age.com/the-five-pillars-of-edge-computing-123485531/
21. Becvar, Z., Plachy, J., Mach, P.: Path selection using handover in mobile networks with cloud-enabled small cells. In: 2014 IEEE 25th Annual International Symposium on Personal, Indoor, and Mobile Radio Communication (PIMRC), pp. 1480–1485. IEEE (2014)
22. Bi, S., Zhang, Y.J.: Computation rate maximization for wireless powered mobile-edge computing with binary computation offloading. IEEE Trans. Wirel. Commun. **17**(6), 4177–4190 (2018)
23. Chen, J., Ran, X.: Deep learning with edge computing: a review. Proc. IEEE **107**(8), 1655–1674 (2019)
24. Chen, M.H., Liang, B., Dong, M.: Joint offloading and resource allocation for computation and communication in mobile cloud with computing access point. In: IEEE INFOCOM 2017-IEEE Conference on Computer Communications, pp. 1–9. IEEE (2017)
25. Chen, S., Xu, H., Liu, D., Hu, B., Wang, H.: A vision of IoT: applications, challenges, and opportunities with china perspective. IEEE Internet Things J. **1**(4), 349–359 (2014)
26. Chen, X., Jiao, L., Li, W., Fu, X.: Efficient multi-user computation offloading for mobile-edge cloud computing. IEEE/ACM Trans. Netw. **24**(5), 2795–2808 (2016)
27. Cicirelli, F., Guerrieri, A., Spezzano, G., Vinci, A., Briante, O., Iera, A., Ruggeri, G.: Edge computing and social internet of things for large-scale smart environments development. IEEE Internet Things J. **5**(4), 2557–2571 (2017)
28. Cong, S., Lakafosis, V., Ammar, M.H., Zegura, E.W.: Serendipity: enabling remote computing among intermittently connected mobile devices. In: ACM Mobihoc (2012)
29. Cui, M., Fei, Y., Liu, Y.: A survey on secure deployment of mobile services in edge computing. Secur. Commun. Netw. **2021** (2021)
30. Deng, M., Tian, H., Fan, B.: Fine-granularity based application offloading policy in cloud-enhanced small cell networks. In: 2016 IEEE International Conference on Communications Workshops (ICC), pp. 638–643. IEEE (2016)
31. Di Valerio, V., Presti, F.L.: Optimal virtual machines allocation in mobile femto-cloud computing: An mdp approach. In: 2014 IEEE Wireless Communications and Networking Conference Workshops (WCNCW), pp. 7–11. IEEE (2014)
32. Dinh, H.T., Lee, C., Niyato, D., Wang, P.: A survey of mobile cloud computing: architecture, applications, and approaches. Wirel. Commun. Mob. comput. **13**(18), 1587–1611 (2013)

33. Dong, L., Satpute, M.N., Shan, J., Liu, B., Yu, Y., Yan, T.: Computation offloading for mobile-edge computing with multi-user. In: 2019 IEEE 39th international conference on distributed computing systems (ICDCS), pp. 841–850. IEEE (2019)
34. Elgendy, I., Muthanna, A., Hammoudeh, M., Shaiba, H.A., Unal, D., Khayyat, M.: Security-aware data offloading and resource allocation for MEC systems: a deep reinforcement learning (2021)
35. Elgendy, I., Zhang, W., Liu, C., Hsu, C.H.: An efficient and secured framework for mobile cloud computing. IEEE Trans. Cloud Comput. (2018)
36. Elgendy, I.A., El-kawkagy, M., Keshk, A.: Improving the performance of mobile applications using cloud computing. In: 2014 9th International Conference on Informatics and Systems, pp. PDC–109. IEEE (2014)
37. Elgendy, I.A., Muthanna, A., Hammoudeh, M., Shaiba, H., Unal, D., Khayyat, M.: Advanced deep learning for resource allocation and security aware data offloading in industrial mobile edge computing. Big Data (2021)
38. Elgendy, I.A., Zhang, W., Tian, Y.C., Li, K.: Resource allocation and computation offloading with data security for mobile edge computing. Future Gener. Comput. Syst. **100**, 531–541 (2019)
39. Elgendy, I.A., Zhang, W.Z., He, H., Gupta, B.B., Abd El-Latif, A.A.: Joint computation offloading and task caching for multi-user and multi-task MEC systems: reinforcement learning-based algorithms. Wirel. Netw. 1–16 (2021)
40. Elgendy, I.A., Zhang, W.Z., Zeng, Y., He, H., Tian, Y.C., Yang, Y.: Efficient and secure multi-user multi-task computation offloading for mobile-edge computing in mobile IoT networks. IEEE Trans. Netw. Serv. Manag. **17**(4), 2410–2422 (2020)
41. Elgendy, M., Herperger, M., Guzsvinecz, T., Lanyi, C.S.: Indoor navigation for people with visual impairment using augmented reality markers. In: 2019 10th IEEE International Conference on Cognitive Infocommunications (CogInfoCom), pp. 425–430. IEEE (2019)
42. Elgendy, M., Sik-Lanyi, C., Kelemen, A.: Making shopping easy for people with visual impairment using mobile assistive technologies. Appl. Sci. **9**(6), 1061 (2019)
43. Elgendy, M., Sik-Lanyi, C., Kelemen, A.: A novel marker detection system for people with visual impairment using the improved tiny-yolov3 model. Comput. Methods Programs Biomed. 106112 (2021)
44. Elgendy, M.A., Shawish, A., Moussa, M.I.: An enhanced version of the MCACC to augment the computing capabilities of mobile devices using cloud computing. Int. Jo. Adv. Comput. Sci. Appl. (IJACSA), Special Issue on Extended Papers from Science and Information Conference. Citeseer (2014)
45. Elgendy, M.A., Shawish, A., Moussa, M.I.: Mcacc: New approach for augmenting the computing capabilities of mobile devices with cloud computing. In: 2014 Science and Information Conference, pp. 79–86. IEEE (2014)
46. Elminaam, D.S.A., Alanezi, F.T., Hosny, K.M.: SMCACC: developing an efficient dynamic secure framework for mobile capabilities augmentation using cloud computing. IEEE Access **7**, 120214–120237 (2019)
47. Farris, I., Taleb, T., Flinck, H., Iera, A.: Providing ultra-short latency to user-centric 5g applications at the mobile network edge. Trans. Emerg. Telecommun. Technol. **29**(4), e3169 (2018)
48. Ghaleb, S.M., Subramaniam, S., Zukarnain, Z.A., Muhammed, A.: Mobility management for IoT: a survey. EURASIP J. Wirel. Commun. Netw. **2016**(1), 165 (2016)
49. Gu, X., Ji, C., Zhang, G.: Energy-optimal latency-constrained application offloading in mobile-edge computing. Sensors **20**(11), 3064 (2020)
50. Guan, L., Ke, X., Song, M., Song, J.: A survey of research on mobile cloud computing. In: 2011 10th IEEE/ACIS International Conference on Computer and Information Science, pp. 387–392. IEEE (2011)
51. Gubbi, J., Buyya, R., Marusic, S., Palaniswami, M.: Internet of things (IoT): a vision, architectural elements, and future directions. Future Gener. Comput. Syst. **29**(7), 1645–1660 (2013)
52. Guo, X., Singh, R., Zhao, T., Niu, Z.: An index based task assignment policy for achieving optimal power-delay tradeoff in edge cloud systems. In: 2016 IEEE International Conference on Communications (ICC), pp. 1–7. IEEE (2016)

53. Gupta, B., Quamara, M.: An overview of internet of things (IoT): architectural aspects, challenges, and protocols. Concurrency Comput.: Pract. Exp. **32**(21), e4946 (2020)
54. Hammad, M., Iliyasu, A.M., Subasi, A., Ho, E.S., Abd El-Latif, A.A.: A multitier deep learning model for arrhythmia detection. IEEE Trans. Instrum. Meas. **70**, 1–9 (2020)
55. Hammad, M., Pławiak, P., Wang, K., Acharya, U.R.: Resnet-attention model for human authentication using ECG signals. Expert Syst. e12547 (2020)
56. Huang, L., Bi, S., Zhang, Y.J.: Deep reinforcement learning for online computation offloading in wireless powered mobile-edge computing networks. IEEE Trans. Mob. Comput. (2020)
57. Huang, L., Feng, X., Feng, A., Huang, Y., Qian, L.P.: Distributed deep learning-based offloading for mobile edge computing networks. Mob. Netw. Appl. 1–8 (2018)
58. Huang, L., Feng, X., Zhang, C., Qian, L., Wu, Y.: Deep reinforcement learning-based joint task offloading and bandwidth allocation for multi-user mobile edge computing. Digit. Commun. Netw. **5**(1), 10–17 (2019)
59. Huang, L., Feng, X., Zhang, L., Qian, L., Wu, Y.: Multi-server multi-user multi-task computation offloading for mobile edge computing networks. Sensors **19**(6), 1446 (2019)
60. Ibrahim, M.H.: Octopus: an edge-fog mutual authentication scheme. IJ Netw. Secur. **18**(6), 1089–1101 (2016)
61. Jiang, C., Cheng, X., Gao, H., Zhou, X., Wan, J.: Toward computation offloading in edge computing: a survey. IEEE Access **7**, 131543–131558 (2019)
62. Kao, Y.H., Krishnamachari, B., Ra, M.R., Bai, F.: Hermes: latency optimal task assignment for resource-constrained mobile computing. IEEE Trans. Mob. Comput. **16**(11), 3056–3069 (2017)
63. Khalili, S., Simeone, O.: Inter-layer per-mobile optimization of cloud mobile computing: a message-passing approach. Trans. Emerg. Telecommun. Technol. **27**(6), 814–827 (2016)
64. Khan, A.U.R., Othman, M., Madani, S.A., Ullah, K.S.: A survey of mobile cloud computing application models. IEEE Commun. Surv. Tut. **16**(1), 393–413 (2013)
65. Khan, U.A., Khalid, W., Saifullah, S.: Energy efficient resource allocation and computation offloading strategy in a uav-enabled secure edge-cloud computing system. In: 2020 IEEE International Conference on Smart Internet of Things (SmartIoT), pp. 58–63. IEEE (2020)
66. Khayyat, M., Alshahrani, A., Alharbi, S., Elgendy, I., Paramonov, A., Koucheryavy, A.: Multi-level service-provisioning-based autonomous vehicle applications. Sustainability **12**(6), 2497 (2020)
67. Khayyat, M., Elgendy, I.A., Muthanna, A., Alshahrani, A.S., Alharbi, S., Koucheryavy, A.: Advanced deep learning-based computational offloading for multilevel vehicular edge-cloud computing networks. IEEE Access **8**, 137052–137062 (2020)
68. Kovachev, D., Cao, Y., Klamma, R.: Mobile cloud computing: a comparison of application models (2011). arXiv preprint arXiv:1107.4940
69. Kumar, K., Lu, Y.H.: Cloud computing for mobile users: can offloading computation save energy? Computer **43**(4), 51–56 (2010)
70. Li, G., He, J., Peng, S., Jia, W., Wang, C., Niu, J., Yu, S.: Energy efficient data collection in large-scale internet of things via computation offloading. IEEE Internet Things J. **6**(3), 4176–4187 (2018)
71. Li, H., Ota, K., Dong, M.: Learning IoT in edge: deep learning for the internet of things with edge computing. IEEE Netw. **32**(1), 96–101 (2018)
72. Li, K., Tao, M., Chen, Z.: Exploiting computation replication for mobile edge computing: A fundamental computation-communication tradeoff study. IEEE Trans. Wirel. Commun. (2020)
73. Li, K.C., Gupta, B.B.: Recent advances in security, privacy, and trust for internet of things (IoT) and cyber-physical systems (CPS) (2020)
74. Li, S., Tao, Y., Qin, X., Liu, L., Zhang, Z., Zhang, P.: Energy-aware mobile edge computation offloading for IoT over heterogenous networks. IEEE Access **7**, 13092–13105 (2019)
75. Lim, W.Y.B., Luong, N.C., Hoang, D.T., Jiao, Y., Liang, Y.C., Yang, Q., Niyato, D., Miao, C.: Federated learning in mobile edge networks: a comprehensive survey. IEEE Commun. Surv. Tutor. (2020)

76. Lin, L., Liao, X., Jin, H., Li, P.: Computation offloading toward edge computing. Proc. IEEE **107**(8), 1584–1607 (2019)

77. Liu, F., Huang, Z., Wang, L.: Energy-efficient collaborative task computation offloading in cloud-assisted edge computing for IoT sensors. Sensors **19**(5), 1105 (2019)

78. Liu, F., Tang, G., Li, Y., Cai, Z., Zhang, X., Zhou, T.: A survey on edge computing systems and tools. Proc. IEEE **107**(8), 1537–1562 (2019)

79. Liu, J., Mao, Y., Zhang, J., Letaief, K.B.: Delay-optimal computation task scheduling for mobile-edge computing systems. In: 2016 IEEE International Symposium on Information Theory (ISIT), pp. 1451–1455. IEEE (2016)

80. Liu, L., Chang, Z., Guo, X., Ristaniemi, T.: Multi-objective optimization for computation offloading in mobile-edge computing. In: 2017 IEEE Symposium on Computers and Communications (ISCC), pp. 832–837. IEEE (2017)

81. Liu, Y., Peng, J., Kang, J., Iliyasu, A.M., Niyato, D., Abd El-Latif, A.A.: A secure federated learning framework for 5G networks. IEEE Wirel. Commun. **27**(4), 24–31 (2020)

82. Mach, P., Becvar, Z.: Cloud-aware power control for cloud-enabled small cells. In: 2014 IEEE Globecom Workshops (GC Wkshps), pp. 1038–1043. IEEE (2014)

83. Mach, P., Becvar, Z.: Cloud-aware power control for real-time application offloading in mobile edge computing. Trans. Emerg. Telecommun. Technol. **27**(5), 648–661 (2016)

84. Mach, P., Becvar, Z.: Mobile edge computing: a survey on architecture and computation offloading. IEEE Commun. Surv. Tutor. **19**(3), 1628–1656 (2017)

85. Mahmoodi, S.E., Uma, R., Subbalakshmi, K.: Optimal joint scheduling and cloud offloading for mobile applications. IEEE Trans. Cloud Comput. (2016)

86. Mao, Y., Hong, W., Wang, H., Li, Q., Zhong, S.: Privacy-preserving computation offloading for parallel deep neural networks training. IEEE Trans. Parall. Distrib. Syst. (2020)

87. Mao, Y., You, C., Zhang, J., Huang, K., Letaief, K.B.: A survey on mobile edge computing: the communication perspective. IEEE Commun. Surv. Tutor. **19**(4), 2322–2358 (2017)

88. Mao, Y., Zhang, J., Letaief, K.B.: Dynamic computation offloading for mobile-edge computing with energy harvesting devices. IEEE J. Sel. Areas Commun. **34**(12), 3590–3605 (2016)

89. Mao, Y., Zhang, J., Letaief, K.B.: Joint task offloading scheduling and transmit power allocation for mobile-edge computing systems. In: 2017 IEEE Wireless Communications and Networking Conference (WCNC), pp. 1–6. IEEE (2017)

90. Mao, Y., Zhang, J., Song, S., Letaief, K.B.: Power-delay tradeoff in multi-user mobile-edge computing systems. In: 2016 IEEE Global Communications Conference (GLOBECOM), pp. 1–6. IEEE (2016)

91. Masud, M., Gaba, G.S., Alqahtani, S., Muhammad, G., Gupta, B., Kumar, P., Ghoneim, A.: A lightweight and robust secure key establishment protocol for internet of medical things in COVID-19 patients care. IEEE Internet Things J. (2020)

92. Mollah, M.B., Azad, M.A.K., Vasilakos, A.: Security and privacy challenges in mobile cloud computing: survey and way ahead. J. Netw. Comput. Appl. **84**, 38–54 (2017)

93. Nadembega, A., Hafid, A.S., Brisebois, R.: Mobility prediction model-based service migration procedure for follow me cloud to support QOS and QOE. In: 2016 IEEE International Conference on Communications (ICC), pp. 1–6. IEEE (2016)

94. Nakkar, M., Altawy, R., Youssef, A.: Lightweight broadcast authentication protocol for edge-based applications. IEEE Internet Things J. **7**(12), 11766–11777 (2020)

95. Ngueilbaye, A., Wang, H., Mahamat, D.A., Elgendy, I.A.: SDLER: stacked dedupe learning for entity resolution in big data era. J. Supercomput. 1–25 (2021)

96. Ning, Z., Dong, P., Kong, X., Xia, F.: A cooperative partial computation offloading scheme for mobile edge computing enabled internet of things. IEEE Internet Things J. **6**(3), 4804–4814 (2018)

97. Noor, T.H., Zeadally, S., Alfazi, A., Sheng, Q.Z.: Mobile cloud computing: challenges and future research directions. J. Netw. Comput. Appl. **115**, 70–85 (2018)

98. Othman, M., Madani, S.A., Khan, S.U., et al.: A survey of mobile cloud computing application models. IEEE Commun. Surv. Tutor. **16**(1), 393–413 (2013)

99. Oueis, J., Strinati, E.C., Barbarossa, S.: Small cell clustering for efficient distributed cloud computing. In: 2014 IEEE 25th Annual International Symposium on Personal, Indoor, and Mobile Radio Communication (PIMRC), pp. 1474–1479. IEEE (2014)

100. Oueis, J., Strinati, E.C., Sardellitti, S., Barbarossa, S.: Small cell clustering for efficient distributed fog computing: a multi-user case. In: 2015 IEEE 82nd Vehicular Technology Conference (VTC2015-Fall), pp. 1–5. IEEE (2015)

101. Pan, J., McElhannon, J.: Future edge cloud and edge computing for internet of things applications. IEEE Internet Things J. **5**(1), 439–449 (2017)

102. Paramonov, A., Muthanna, A., Aboulola, O.I., Elgendy, I.A., Alharbey, R., Tonkikh, E., Koucheryavy, A.: Beyond 5g network architecture study: fractal properties of access network. Appl. Sci. **10**(20), 7191 (2020)

103. Plachy, J., Becvar, Z., Mach, P.: Path selection enabling user mobility and efficient distribution of data for computation at the edge of mobile network. Comput. Netw. **108**, 357–370 (2016)

104. Plachy, J., Becvar, Z., Strinati, E.C.: Dynamic resource allocation exploiting mobility prediction in mobile edge computing. In: 2016 IEEE 27th Annual International Symposium on Personal, Indoor, and Mobile Radio Communications (PIMRC), pp. 1–6. IEEE (2016)

105. Qi, Q., Tao, F.: A smart manufacturing service system based on edge computing, fog computing, and cloud computing. IEEE Access **7**, 86769–86777 (2019)

106. Rafique, W., Qi, L., Yaqoob, I., Imran, M., ur Rasool, R., Dou, W.: Complementing iot services through software defined networking and edge computing: a comprehensive survey. IEEE Commun. Surv. Tutor. (2020)

107. Rahmani, A.M., Gia, T.N., Negash, B., Anzanpour, A., Azimi, I., Jiang, M., Liljeberg, P.: Exploiting smart e-health gateways at the edge of healthcare internet-of-things: a fog computing approach. Future Gener. Comput. Syst. **78**, 641–658 (2018)

108. Ren, J., Yu, G., Cai, Y., He, Y.: Latency optimization for resource allocation in mobile-edge computation offloading. IEEE Trans. Wirel. Commun. **17**(8), 5506–5519 (2018)

109. Rodrigues, T.G., Suto, K., Nishiyama, H., Kato, N., Temma, K.: Cloudlets activation scheme for scalable mobile edge computing with transmission power control and virtual machine migration. IEEE Trans. Comput. **67**(9), 1287–1300 (2018)

110. Roman, R., Lopez, J., Mambo, M.: Mobile edge computing, fog et al.: a survey and analysis of security threats and challenges. Future Gener. Comput. Syste. **78**, 680–698 (2018)

111. Saleem, U., Liu, Y., Jangsher, S., Tao, X., Li, Y.: Latency minimization for D2D-enabled partial computation offloading in mobile edge computing. IEEE Trans. Veh. Technol. **69**(4), 4472–4486 (2020)

112. Sanaei, Z., Abolfazli, S., Gani, A., Buyya, R.: Heterogeneity in mobile cloud computing: taxonomy and open challenges. IEEE Commun. Surv. Tutor. **16**(1), 369–392 (2013)

113. Satyanarayanan, M., Bahl, P., Caceres, R., Davies, N.: The case for VM-based cloudlets in mobile computing. IEEE Pervasive Comput. **8**(4), 14–23 (2009)

114. Schneider, M., Rambach, J., Stricker, D.: Augmented reality based on edge computing using the example of remote live support. In: 2017 IEEE International Conference on Industrial Technology (ICIT), pp. 1277–1282. IEEE (2017)

115. Schulz, P., Matthe, M., Klessig, H., Simsek, M., Fettweis, G., Ansari, J., Ashraf, S.A., Almeroth, B., Voigt, J., Riedel, I., et al.: Latency critical IoT applications in 5G: perspective on the design of radio interface and network architecture. IEEE Commun. Mag. **55**(2), 70–78 (2017)

116. Secci, S., Raad, P., Gallard, P.: Linking virtual machine mobility to user mobility. IEEE Trans. Netw. Serv. Manag. **13**(4), 927–940 (2016)

117. Sedik, A., Hammad, M., Abd El-Samie, F.E., Gupta, B.B., Abd El-Latif, A.A.: Efficient deep learning approach for augmented detection of coronavirus disease. Neural Comput. Appl. 1–18 (2021)

118. Sheng, Z., Mahapatra, C., Leung, V.C., Chen, M., Sahu, P.K.: Energy efficient cooperative computing in mobile wireless sensor networks. IEEE Trans. Cloud Comput. **6**(1), 114–126 (2015)

119. Stergiou, C., Psannis, K.E., Kim, B.G., Gupta, B.: Secure integration of IoT and cloud computing. Future Gener. Comput. Syst. **78**, 964–975 (2018)
120. Stergiou, C.L., Psannis, K.E., Gupta, B.B.: Iot-based big data secure management in the fog over a 6g wireless network. IEEE Internet Things J. (2020)
121. Sun, X., Ansari, N.: Primal: Profit maximization avatar placement for mobile edge computing. In: 2016 IEEE International Conference on Communications (ICC), pp. 1–6. IEEE (2016)
122. Tanzil, S.S., Gharehshiran, O.N., Krishnamurthy, V.: Femto-cloud formation: a coalitional game-theoretic approach. In: 2015 IEEE Global Communications Conference (GLOBE-COM), pp. 1–6. IEEE (2015)
123. Tao, F., Zhang, M., Nee, A.Y.C.: Digital Twin Driven Smart Manufacturing. Academic Press (2019)
124. Tawalbeh, L., Al-Qassas, R.S., Darwazeh, N.S., Jararweh, Y., AlDosari, F.: Secure and efficient cloud computing framework. In: 2015 International Conference on Cloud and Autonomic Computing, pp. 291–295. IEEE (2015)
125. Vallina-Rodriguez, N., Crowcroft, J.: Energy management techniques in modern mobile handsets. IEEE Commun. Surv. Tutor. **15**(1), 179–198 (2013). https://doi.org/10.1109/SURV.2012.021312.00045
126. Wang, C., He, Y., Yu, F.R., Chen, Q., Tang, L.: Integration of networking, caching, and computing in wireless systems: a survey, some research issues, and challenges. IEEE Commun. Surv. Tutor. **20**(1), 7–38 (2017)
127. Wang, F., Xu, J., Cui, S.: Optimal energy allocation and task offloading policy for wireless powered mobile edge computing systems. IEEE Trans. Wirel. Commun. **19**(4), 2443–2459 (2020)
128. Wang, F., Xu, Y., Zhu, L., Du, X., Guizani, M.: Lamanco: a lightweight anonymous mutual authentication scheme for n-times computing offloading in iot. IEEE Internet Things J. **6**(3), 4462–4471 (2019)
129. Wang, J., Lv, T., Huang, P., Mathiopoulos, P.T.: Mobility-aware partial computation offloading in vehicular networks: a deep reinforcement learning based scheme. China Commun. **17**(10), 31–49 (2020)
130. Wang, J., Pan, J., Esposito, F., Calyam, P., Yang, Z., Mohapatra, P.: Edge cloud offloading algorithms: issues, methods, and perspectives. ACM Comput. Surv. (CSUR) **52**(1), 1–23 (2019)
131. Wang, S., Urgaonkar, R., He, T., Chan, K., Zafer, M., Leung, K.K.: Dynamic service placement for mobile micro-clouds with predicted future costs. IEEE Trans. Parall. Distrib. Syst. **28**(4), 1002–1016 (2016)
132. Wang, S., Zafer, M., Leung, K.K.: Online placement of multi-component applications in edge computing environments. IEEE Access **5**, 2514–2533 (2017)
133. Wang, S., Zhang, X., Zhang, Y., Wang, L., Yang, J., Wang, W.: A survey on mobile edge networks: convergence of computing, caching and communications. IEEE Access **5**, 6757–6779 (2017)
134. Wang, X., Han, Y., Leung, V.C., Niyato, D., Yan, X., Chen, X.: Convergence of edge computing and deep learning: a comprehensive survey. IEEE Commun. Surv. Tutor. **22**(2), 869–904 (2020)
135. Wang, Y., Sheng, M., Wang, X., Wang, L., Li, J.: Mobile-edge computing: partial computation offloading using dynamic voltage scaling. IEEE Trans. Commun. **64**(10), 4268–4282 (2016)
136. Xiao, Y., Jia, Y., Liu, C., Cheng, X., Yu, J., Lv, W.: Edge computing security: state of the art and challenges. Proc. IEEE **107**(8), 1608–1631 (2019)
137. Xu, X., Gu, R., Dai, F., Qi, L., Wan, S.: Multi-objective computation offloading for internet of vehicles in cloud-edge computing. Wirel. Netw. 1–19 (2019)
138. Xu, X., Xue, Y., Qi, L., Yuan, Y., Zhang, X., Umer, T., Wan, S.: An edge computing-enabled computation offloading method with privacy preservation for internet of connected vehicles. Future Gener. Comput. Syst. **96**, 89–100 (2019)
139. Yadav, R., Zhang, W., Kaiwartya, O., Singh, P.R., Elgendy, I.A., Tian, Y.C.: Adaptive energy-aware algorithms for minimizing energy consumption and SLA violation in cloud computing. IEEE Access **6**, 55923–55936 (2018)

140. Yaqoob, I., Ahmed, E., Gani, A., Mokhtar, S., Imran, M., Guizani, S.: Mobile ad hoc cloud: a survey. Wirel. Commun. Mob. Comput. **16**(16), 2572–2589 (2016)
141. Yi, S., Qin, Z., Li, Q.: Security and privacy issues of fog computing: a survey. In: International Conference on Wireless Algorithms, Systems, and Applications, pp. 685–695. Springer (2015)
142. Yu, F., Chen, H., Xu, J.: DMPO: dynamic mobility-aware partial offloading in mobile edge computing. Future Gener. Comput. Syst. **89**, 722–735 (2018)
143. Yu, Y., Zhang, J., Letaief, K.B.: Joint subcarrier and cpu time allocation for mobile edge computing. In: 2016 IEEE Global Communications Conference (GLOBECOM), pp. 1–6. IEEE (2016)
144. Zaghloul, A., Zhang, T., Hou, H., Amin, M., Abd El-Latif, A.A., Abd El-Wahab, M.S.: A block encryption scheme for secure still visual data based on one-way coupled map lattice. Int. J. Secur. Appl. **8**(4), 89–100 (2014)
145. Zhang, C., Patras, P., Haddadi, H.: Deep learning in mobile and wireless networking: a survey. IEEE Commun. Surv. Tutor. (2019)
146. Zhang, K., Mao, Y., Leng, S., Zhao, Q., Li, L., Peng, X., Pan, L., Maharjan, S., Zhang, Y.: Energy-efficient offloading for mobile edge computing in 5G heterogeneous networks. IEEE Access **4**, 5896–5907 (2016)
147. Zhang, T.J., Manhrawy, I., Abdo, A., Abd El-Latif, A., Rhouma, R.: Cryptanalysis of elementary cellular automata based image encryption. In: Advanced Materials Research, vol. 981, pp. 372–375. Trans Tech Publ (2014)
148. Zhang, W.Z., Elgendy, I.A., Hammad, M., Iliyasu, A.M., Du, X., Guizani, M., Abd El-Latif, A.A.: Secure and optimized load balancing for multi-tier IoT and edge-cloud computing systems. IEEE Internet Things J. (2020)
149. Zhang, X., Mao, Y., Zhang, J., Letaief, K.B.: Multi-objective resource allocation for mobile edge computing systems. In: 2017 IEEE 28th Annual International Symposium on Personal, Indoor, and Mobile Radio Communications (PIMRC), pp. 1–5. IEEE (2017)
150. Zhao, T., Zhou, S., Guo, X., Zhao, Y., Niu, Z.: A cooperative scheduling scheme of local cloud and internet cloud for delay-aware mobile cloud computing. In: 2015 IEEE Globecom Workshops (GC Wkshps), pp. 1–6. IEEE (2015)
151. Zhao, T., Zhou, S., Song, L., Jiang, Z., Guo, X., Niu, Z.: Energy-optimal and delay-bounded computation offloading in mobile edge computing with heterogeneous clouds. China Commun. **17**(5), 191–210 (2020)
152. Zheng, X., Li, M., Chen, Y., Guo, J., Alam, M., Hu, W.: Blockchain-based secure computation offloading in vehicular networks. IEEE Trans. Intell. Transp. Syst. (2020)
153. Zheng, Y., Lu, R., Mamun, M.: Privacy-preserving computation offloading for time-series activities classification in ehealthcare. In: ICC 2020-2020 IEEE International Conference on Communications (ICC), pp. 1–6. IEEE (2020)
154. Zhu, X., Li, J., Liu, Z., Yang, F.: Location deployment of depots and resource relocation for connected car-sharing systems through mobile edge computing. Int. J. Distrib. Sens. Netw. **13**(6), 1420–1435 (2017). http://orcid.org/1550147717711621

Security and Interoperability Issues with Internet of Things (IoT) in Healthcare Industry: A Survey

Eman M. Abounassar, Passent El-Kafrawy, and Ahmed A. Abd El-Latif

Abstract Recently, public healthcare systems become one of the most pivotal parts in our daily life. Resulting in an insane increase in Medical data like medical images and patient information. Having huge amount of data requires more computational power for efficient data management. In addition, data security, privacy and trustworthy have to be maintained and guaranteed. Most medical information in the last years aggregated data from a lot of devices, smart chips, tiny sensors and wearable devices. Those devices are connected through the internet, thus called Internet of Things (IoT). These devices and objects are considered components of the health care technology. Unfortunately, health technologies still face aplenty of privacy preserving, trust issues and a lot of other security problems when transferred through networks. In this work, we will present a security and privacy mechanisms and tools dealing with patient medical Data (information, images) given from smart devices and sensors upon internet of healthcare things (IoHT). Presenting some of the future security trends to enhancing the trustworthy and satisfaction level for healthcare ecosystems.

Keywords IoT · Healthcare · Security · Privacy · Trustworthy · IoHT

1 Introduction

In Health-IoT environments, IoT targets the integration between the physical world and virtual one, where things can exchange information and communicate with each other [1, 2]. This association between smart devices through IoT makes conceivably an advanced portrayal of this present reality. Through which shifted fields of intelligent applications in different industries and ventures can be developed. Applications in the medical domain, smart cities, waste management, traffic congestion,

E. M. Abounassar (✉) · P. El-Kafrawy · A. A. Abd El-Latif
Mathematics and Computer Science Department, Menoufia University, Shebin El-Kom 32511, Egypt

P. El-Kafrawy
School of Information Technology and Computer Science, Nile University, Giza 12588, Egypt

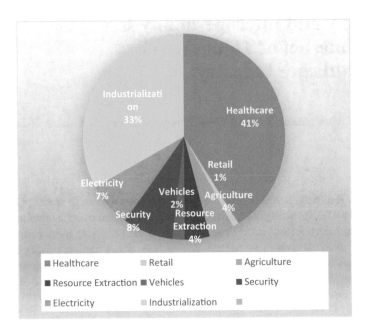

Fig. 1 IoT applications

security, emergency services, 4G industry, healthcare and transportation can benefit from such advancement [3–5]. Healthcare and biomedical domains are considered one of the most appealing applications for the IoT-environment as shown in Fig. 1. Enhancing healthcare services and developing smart biomedical systems is one of the most inspiring objectives in the coming era [2, 3, 6].

Various healthcare applications have been highlighted in Fig. 2. It can be noted that notions from previous parts that are proposed by researchers, are implemented in real end to end applications that serve end users, i.e. patients and physicians.

2 IoT Background

Before delving into the concepts and technologies behind IoT, we will first assess 'What is IoT-Domain, what is its use and how does it work?' The nature and abilities of IoT have made it very popular as IoT industries have enhanced the devices and their applications in all domains.

The IoT is a wide connection of millions of devices (objects) that can access the internet to share and transfer huge amount of data without any human interference. These objects could be computing devices, mobile devices, mechanical devices, or medical devices that include a very small chip that allows it to connect to the internet either on wired or wireless way.

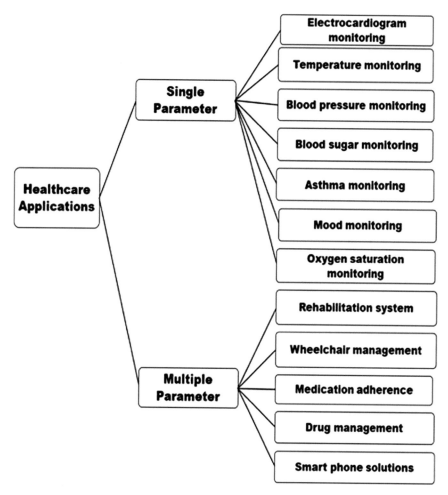

Fig. 2 Healthcare application list

2.1 IoT Vision

The IoT vision developed by many stakeholders based on their interests and usage. The vision was expanded to include means of connectivity to allow either physical objects or virtual objects to be connected anytime, anywhere and to anything using communication methodologies for example: Radio-Frequency Identification (RFID), wired, wireless networks, Bluetooth or Global Positioning System (GPS) [7, 8]. With this vision, the IoT field became bigger in a large network of distributed smart objects besides computer networks to reach common goals [9, 10].

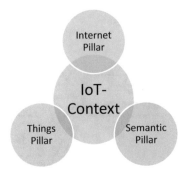

Fig. 3 The main pillars for IoT-context [11]

IoT vision is divided into three pillars as shown in Fig. 3:

1. Things Oriented pillar
2. Internet Oriented pillar
3. Semantic Oriented pillar.

- *Things Oriented Pillar*: refers to any object that can be monitored and tracked by tagging it using RFID tags and identifies these objects by RFID transponders or readers or just using sensors [1, 9].
- *Internet Oriented Pillar*: is constrained on the need of having smart objects linked together within the massive network (internet), by having features of IP protocols. This is one of the prime protocols being used to internet things all over the world. The sensor-based object can be altered for readability and can be identified peer-lessly; then their characteristics can be persistently tracked and monitored. This marks the basis of smart embedded objects; assumed to be as a microcomputer, having computing resources [12].
- *Semantic Oriented Pillar*: is powered by the truth that the huge numbers of sensors communicated with the internet will collect and transfer huge amount of data. Therefore, there will be an innumerous amount of redundant information that need to be handled meaningfully. The collected data have to be processed, handled in readable style for better exemplification using semantic technologies [10, 12, 13].

2.2 IoT Architecture

IoT as a technology majorly is constructed of four principle layers as shown in Fig. 4.

- Sensors/Devices layer
  The Objects or devices/perception layer is the physical layer that includes the sensors actuators for sensing some parameters such as location, motion, tempera-ture or humidity. These data are transferred to the next layer using wireless sensor network (WSN), RFID or any other communication technology [9, 14].

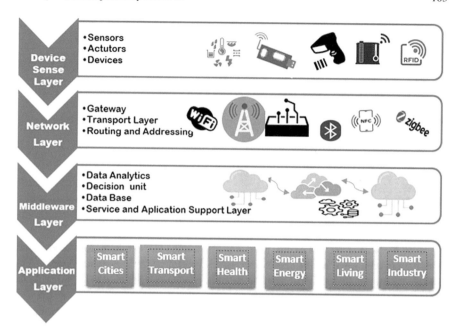

Fig. 4 IoT- Environment layers

- Gateways and Networks layer
  The IoT gateway layer is the carrier layer that takes the responsibility for routing the received data from the Sensors/Devices layer to the middleware layer. Sensor networks use communication methodologies like Wi-Fi, Bluetooth (BT), RFID or ZigBee.
- Middleware/Cloud or Management Service Layer
  The middleware layer is the brain of IoT construction that includes some of data processing software. That process the data coming from the first layer (Sensors/Devices layer) then give intelligence to these data. This layer is liable for adding some sort of security to the data by performing encryption and decryption. The proposed model will target this layer to increase security level with an acceptable level of interoperability.
- Application Layer
  The Application layer is on the top and the end of the IoT construction layers. This layer is liable for giving meaning of the collected data and delivering it to application specific services to the end users (Consumers) for example: smart homes, security systems, and smart health [1, 14].

Fig. 5 The six key elements for IoT [15]

2.3 IoT Elements

After explaining IoT architecture as a high level of IoT system, we have to go deeply in IoT technology and explain its main six elements that deliver all of IoT functionality. These elements are show below in Fig. 5.

Identification is used to identify the services names and ensuring that the services match their demand. Ubiquitous Codes (UCode) and Electronic Product Codes (EPC) are two examples of IoT Identification methods [7–9, 16]. *Sensing* is responsible for collecting the data from objects within a particular network then transfers the data to the Databases (DBs), data warehouse or the cloud. RFID is the *communication* technology that uses the radio waves as a media to send the digital data in RFID tags and capture these data by a reader. Bluetooth uses short-wavelength radio that allows devices to communicate within short distance with minimum power consumption [17–20].

The *computation* element includes two items: Hardware that host the IoT platforms and applications. In addition to the software that includes IoT OS and applications which act as the brain. *Services* can be classified under four categories: Information Aggregation, Collaborative-Aware, Identity-related and Ubiquitous Services. *Semantics* is an added value in IoT especially on machine-interpretable creation as it provides a smarter way to extract knowledge from different devices to supply the needed services in the IoT domain. Hence, semantics is considered the brain of IoT as it delivers the requests to the correct resources [21–23] due to its deep understanding of the data. To execute semantics on IoT-Context, a high level of ontology need to be reached by utilizing technologies like Resource Description Framework (RDF) and Web Ontology Language (OWL). Also, the Efficient XML Interchange (EXI) format is recommended by the World Wide Web Consortium (W3C) in 2011 [24].

2.4 IoT Applications (Industries)

On the highest level of the IoT architecture there are the applications that deliver all needed functionalities to the end user. Although, this level is not a part of middleware, it advocates all middleware functionalities using service composition technologies and web service standards protocols. These applications can do distributed systems and applications integration [1, 9].

Fig. 6 Internet of things
application domains

Current environments contain a lot of objects and devices with simple intelligence without any interconnection or communication most of the time. Adding interconnectivity or communication facility between them means deploying a lot of applications in different environments and domains [14, 25, 26]. These domains are listed below and shown in Fig. 6:

- Transportation domain.
- Healthcare domain.
- Smart home domain.
- Agriculture domain.
- Schools domain.
- Markets domain.

3 Healthcare Based IoT and Benefits

Recently, the Internet of Things (IoT) influence become prevalent and dominant. Various devices and sensors use the internet to gather data. Consequently, these devices became smarter and more autonomous. Due to technology enhancement, the smart devices became flexible in use, able to communicate and transfer information which affected human life.

IoT can be effectively used in the Healthcare field by supplying it with various computing systems that depends on the communication of devices and applications. Such systems connect health providers with patients allowing them to continuously track, monitor, diagnose and store important statistics and medical information.

Healthcare based IoT brings a lot of benefits that improves the efficiency and the accuracy of medical treatment. For example, it can be used in saving a lot of human's lives by placing internet-connected medical devices that send real time and simultaneously monitor and report their case to the doctors. These devices can collect some of the patient data like blood sugar level, Electrocardiogram sensors (ECGs), and blood pressure level. Then transferred to the Hospital Information System (HIS) to be shared with doctor. Decisions can be taken correctly and immediately preserving the patient's health without any complications that are difficult to treat later [3, 27, 33–37].

Below some of other benefits as well for the use of IoT in healthcare systems:

- Reduces the distance barriers by eliminating the physically attendance on medical facilities
- Rapid medical care especially on medical contingency and natural disasters.
- Reduce the documentation and make work paperless.
- Provide similar and equitable healthcare for all by eliminating geographical barriers.
- Increase access to different healthcare services providers.
- Best communication for both healthcare doctors and specialists that happens at the same time as they are virtually in the same patient room sharing the diagnosis.

3.1 IoT Healthcare Technology

Healthcare systems can be enhanced by connecting devices, such as various tiny sensors, to construct an IoT network devoted to healthcare assessment. Such network allows for remote health monitoring and automatic critical case detection; for elderly care, fitness programs and chronic diseases (heart failure, respiratory diseases, hypertension or diabetes) [38]. IoT healthcare systems shall decrease time and cost of services, increase safety and quality of healthcare, with a continuous medical care and higher user's experience [36, 37, 39]. IoT-healthcare applications have very important characteristics due the sensitivity of the data. Information is generated in huge volume which congests the network and consumes great processing power. Health-IoT environments are significant and essential to set up safely as it requires high security, protection and privacy techniques to avoid any type of security break and vulnerability. We shall study methods that can increase security and protection in e-Health. To have proficient e-Health arrangement, it is significant to survey current solutions proposed in most recent system or model for e-Health.

3.2 Types of Healthcare Devices/Gadgets

There are four classes of healthcare devices: Medical Wearable gadgets, Medical Embedded gadgets, Wearable Health Monitoring gadgets, and Stationary Medical

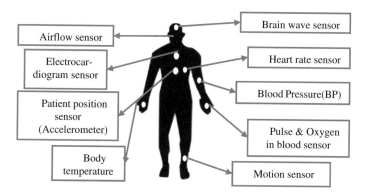

Fig. 7 Sensory wearables in human boby

gadgets. These gadgets can be installed into cloths or worn on the bodies like accessories. These electronic devices are connected with sophisticated features like real-time, feedback, wireless data transmission, and alerting mechanisms built inside the devices. Furthermore, it has the ability to provide important data to healthcare-caregivers such as breathing patterns, blood glucose levels, heart rate, respiration rate and blood pressure as shown in Fig. 7.

Healthcare sensitive data is a profitable target for attackers. Therefore, securing health information is the primary motivation of healthcare providers. Health technologies yet face many privacy preserving risks and security problems while data is transferred over the networks [27, 40].

3.3 Healthcare Concepts

The fusion of healthcare and IoT-network upon cloud computing has been revolutionized, in many concepts by the research community. Every concept supports a collection of healthcare applications, as highlighted in Fig. 8. It is difficult to provide a common or general clarification for the idea of healthcare IoT and cloud computing. Several fundamental concepts will be highlight in this section as a common base for solutions for different applications. Continuous development of healthcare systems generated new concepts. Therefore, they will become the base for medical services applications.

Fig. 8 Healthcare
conceptions

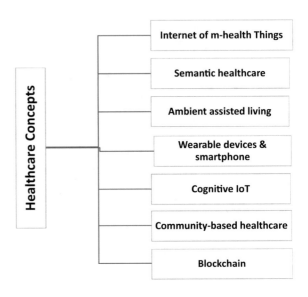

4 Medical Internet of Things (M-IoT)

Recently, M-IoT became an attractive domain in healthcare which refers to using mobile computing, medical sensors, and cloud computing for real time monitoring of patient's vitals [41, 42]. Utilizing communication techniques to transfer the data to a cloud computing framework. Practitioners can get back the data to observe, diagnose, and treat patients efficiently and on time. Therefore, in the future, it has the potency to be the basis for IoT and cloud computing in healthcare applications as it supplies fully mobility and connected features. The challenges that guarantees system security and user's privacy for the M-health applications were mentioned in [43]. The authors suggested different methods to rise the confidentiality of patient's data that includes: physical safeguards, technical safeguards, audit reports, technical policies, and network security.

4.1 Semantic Healthcare

Healthcare systems uses the semantic applications and ontologies to manage and store massive amounts of data [44, 45]. The presence of semantics and ontologies allowed semantic interoperability between various wearable devices in the healthcare systems. Moreover, authors in [44] introduced semantic healthcare framework over IoT environment to support communication among numerous IoT devices. They have designed a lightweight semantic annotation model for data in which makes it

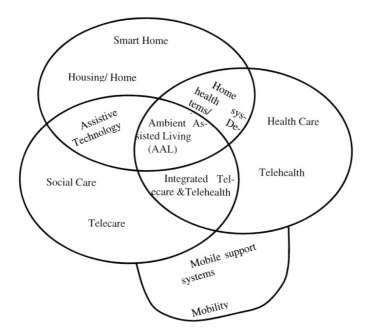

Fig. 9 Structure of ambient assisted living (AAL)

semantically meaningful. Issues such as semantic modeling and semantic interoperability for big-data in heterogeneous IoT infrastructure for healthcare have been discussed in [3, 46, 47].

4.2 Ambient Assisted Living (AAL)

Ambient assisted living technology (AAL) is a sub-area of ambient intelligence. It distributes smart objects in the ambience to supply help and assistance for elders to live independently as shown in Fig. 9. Nowadays, there are a lot of researches related to AAL and is increasing significantly regarding healthcare gadgets availability and sensors technologies development.

4.3 Wearable Devices and Smartphone

Now a days, wearable technology is a booming topic besides being IoT distinctive characteristic. In healthcare, wearable gadgets minimize the costs and fetch many advantages to health professionals and patients. Wearable gadgets can be anything for example; smart wristbands, smart watches, shoes, T-shirts, caps, smart headbands and digital eyeglasses [27, 28, 44]. Numerous sensors are merged into smart devices to simplify the ability of collecting user's health condition or surrounding ambience, then uploading these data to a database or fog layer for real-time processing. Smartphones support these wearable objects using their computing power to transfer the gathered information to the cloud-based framework for analyzing, processing and storing these data. By the end, analyzed data is visualized by smartphone health application. Authors in [28] introduced a framework of IoT healthcare that contained various wearable devices that collected and analyzeed sensors data from time to time.

4.4 Cognitive IoT (CIoT)

Cognitive IoT based Healthcare allows sensors to be more smart. As a new concept of IoT, cognitive computing designates intelligent devices (objects) that can simulate human brain in resolving issues. The CIoT applications improve sensors processing and adapting with the surrounding environment in an automatic manner. CIoT supports data analytic and detects hidden patterns in big data. In [28] authors introduced CIoT framework using message-oriented middleware and semantic representation.

4.5 Community-Based Healthcare

Community-based healthcare concept means constructing a network that covers each portion of a local community. Community-based healthcare uses cloud computing and IoT in healthcare framework that serve countryside or residential suburb. Many networks can be correlated together to perform a cooperative network environment.

4.6 Blockchain

Not only Data fragmentation is the critical issue that prevents data sharing between healthcare providers but also the hard security needs and trust case. All of these issues have to be addressed and studied carefully to take the advantage of healthcare components. Blockchain technology eliminated these problems and presented a radical penetration to fix the data segmentation issue. One advantage of Blockchain is helping healthcare communities to share the sensitive medical data in a secure way. This is done as Blockchain consists of three basic components to provide the security and the flexibility for data sharing as following:

- *Fixed ledger* that guarantees that once a record in a ledger is stored, it will not be modified, and each process is verified regarding to pre-defined policies.

- *Distributed technology*, Blockchain operated simultaneously by various computers which means there is no single point of failure wherever records or digital assets are hacked or compromised.
- Data exchange logic and rules with resilient techniques for *smart contracts*. The rules are used in healthcare to manage the identities and assign distinct permissions for different Electric Medical Records (EMRs). Blockchain guarantee allowing only doctors to access the EMR profiles. There are many projects that uses the Blockchain to handle EMR, medicine prescriptions, payment distribution, pharmaceutical supply chain and clinical pathways.

5 IoT Issues

The IoT-architectures represent constantly growing network of several diversified things or components. Diversified objects in the IoT-System will need to be communicated in many cases, autonomously in order to offer types of services and functionality. Several issues raised due to the high demand of IoT devices in healthcare systems.

5.1 IoT Interoperability Issues

Semantic web methodologies like ontologies can therefore uphold in tending to gadget heterogeneity in IoT. Things/components could be produced by distinct manufacturers that do not essentially follow the mutual standardization. Moreover, devices often work using a diversity of communication mechanisms. They do not usually connect IoT devices with Internet like typical computers usually do. To accomplish collaboration between and with the real-world objects and provide seamless communication among such heterogeneous devices, at anytime and anywhere in future, we want to interact with interoperable solutions. For that, semantic technologies uphold interoperability among devices and simplify actual data access and utilization. The techniques that use non-interoperable communication lead to rise of complication in communicating among the system components and interpreting their data and services [29, 30].

Interoperability is defined as the ability of two or more implementations of components or systems from distinct manufacturers to share, reuse, and exchange data. They use information from each other by depending on each other's services as specified by a mutual criterion [31]. Figure 10 clarifies an interoperability scenario in the IoT-context. In which, many devices attempting to interconnect through them using distinct communication methods (Bluetooth, Wi-Fi, ZigBee, ZWave, RFID, etc.).

To solve these cases, smart gateway or IoT cloud can be used to provide IoT through them [3]. The interoperability has provided numerous challenges such as information gathering, data exchanging, and using the information to understand

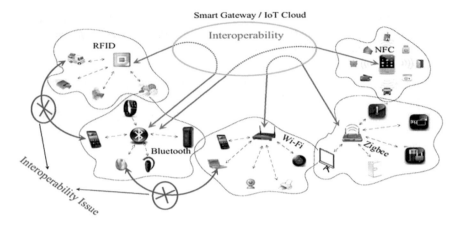

Fig. 10 Scenario of the interoperability in the IoT-environment [48]

and process it. Various categories of interoperability issues are appearing in IoT-context [30, 33]. Different levels of IoT problems which needs to be addressed in supporting seamless and heterogeneous communications in the IoT are presented in Fig. 11.

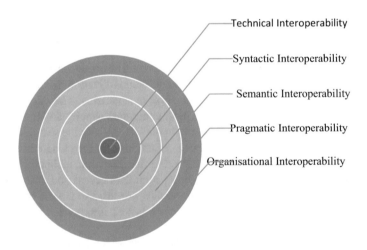

Fig. 11 The interoperability issue levels

5.2 Technical Interoperability

This type commonly concerned with hardware/software components, systems and platforms that enable Machine-to-Machine (M2M) communication to run. Technical interoperability is often based on communication technologies, protocols and the infrastructure needed for these protocols to work. This level ensures common understanding of signals and provides only low-level integration [31].

5.3 Syntactic Interoperability

This type usually is concerned with data structures and formats (e.g.HTML, JSON or XML). Additionally, the carried messages by communication protocols need to have an explicit syntax and encoding, even if it is only in the form of bit-tables. This layer ensures the common understanding of symbols.

5.4 Semantic Interoperability

Semantic interoperability issue commonly concerned with the content meaning, which can carry more than one meaning. This type is predominantly worried about the human instead of M2M interpretation of the content. Semantic interoperability issues provide the meaning of the information which is being exchanged between people [22, 32, 33]. Providing interoperability in software will understand data semantically to permit unambiguous comprehension of the substance. Along these lines, this layer guarantees the normal comprehension of terms.

5.5 Pragmatic Interoperability

Pragmatic or Realistic Interoperability is accomplished when information is utilized, or the setting of its application is known by the worked together frameworks. This layer puts the "word" it means into setting, it guarantees the normal comprehension of the utilization of terms. In this layer, the interoperating systems well-know about methods and procedures that every system use [30].

5.6   Organizational Interoperability

This layer is based on the success of previous stages (technical, syntactical, semantic and pragmatic interoperability). Through which, organizational interoperability level is able to achieve efficient communication and transmit meaningful information among the organizations or companies. This is not accomplished if they are using distinct information systems through widely distinct infrastructures, geographic regions and cultures.

5.7   IoT and Security Issues

The IoT raised another degree of outsourcing but there are some worries about service scalability, accessibility, price structure, response time and licensed innovation proprietorship etc. Security is a main worry when information is transferred among the Web for private organizations and VPN tunnels. The Public authority guidelines, for example, Health Insurance Portability and Accountability (HIPA) Act can be applied as security measures when moving information between global lines. [2, 49].

Many researches focused on security enhancement, privacy preserving techniques, authentication, access policy methods and anonymization to reach satisfaction level for confidentiality to achieve security in IoT network. There are four levels based IoT layers which are listed below and shown in details in Fig. 12 [35].

1. *Sensor data integrity:* through device sense layer,
2. *Authentication:* through network and gateway layer,
3. *Privacy preserving:* the most security measure to be achieved over cloud or middleware layer,

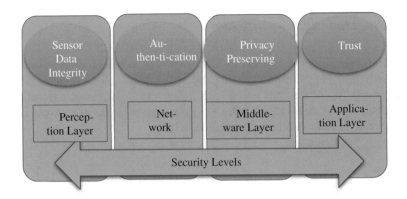

Fig. 12 Security levels based IoT layers

4. *Trustworthy:* consider last level of security is to achieve the trust between members on the IoT network.

The main idea of IoT security tasks is to guarantee a proper protection on application level like Distributed Denial of Service (DDoS) attack mitigation. It must also combine measures to authorize the identity of entities that request accessing to any information including multi-factor confirmation.

6   Related Work

Because of different capabilities and communication protocols, IoT interoperability is still a complex practical and scientific problem. In the first level of interoperability, technical interoperability is considered the foremost fundamental form of interoperability. But the major issue is not using different protocols. There exist proofed solutions to protocols interoperability by using applications gateways. Also, syntactic interoperability stage are used based on the matching between various domains. This is difficult, but is explicit, as the source and the destination are fixed, and the requirements are clear to define the matching between the source and the destination. Nevertheless, semantic interoperability mapping still problematic, since the target is has no standard meaning (ambiguity in the meanings of data). To solve this problem, current researches are typically based their work on using ontologies. Semantic Sensor Network (SSN) is an example of semantic ontology (explained above) that describes networked sensors and their concepts by defining a small set of classes.

Xiao and Guo [29] suggested an approach to resolve the discovery and interaction among various IoT devices. They resolved IoT device of a user context and a device of another context interoperability issue by offering UIF Framework. In which sets of heterogeneous real-devices are capable to be transformed to a set of mutual devices and virtual devices. Any device is both syntactically and semantically transformable between their representations of different contexts. Black stock et al. in [50] in 2014, achieved integrity-based web services through IoT hubs to assemble data utilized by web protocols. Their architecture contains four stages:

1. IoT Core which is concerned with the things and its metadata, which named REST-full web services.
2. The IoT Model which is concerned with the development of the adapters and other integration tools.
3. The IoT Hub which is responsible for the agreement on implementation issues.
4. IoT Profiles which is concerned with the impact of the semantics of things and their related exposed data on a hub.

Desai [22] suggested the perception of Semantic Gateway as Service (SGS) as a bridge between nodes and IoT services in order to supply interoperability using communication and data standards, a node in the topology is the link to the gateway using (CoAP, XMPP or MQTT protocol). The translation between them used multi-protocol proxy. Using SSN (invented by W3C), gateway is the place where data is semantically annotated. Semantic Gateway Bridges are considered as low level raw-sensor information with knowledge centric application services, by facilitating interoperability at messaging protocol and data forming level.

Strassner et al. [33],improved the interoperability by depicting an architecture based on semantics. It merged data models and ontologies to achieve semantic interoperability between IoT entities. It also guarantee that no lost or alteration on the significance of terms and items in a single device (gadget) or framework when swapped or utilized by another gadgets or framework. Unfortunately, they did not take security as an important issue.

Serrano et al. in [51] proposed IoT stack "developed by FP7 OpenIoT" project and its applicable through numerous use cases and executed demos that enabled IoT data interoperability. The OpenIoT platform LSM-Light layer is capable to supply access to data from the sensors.

Androcec et al. [52] achieved a service-level interoperability by suggesting IoT-interoperability framework based on semantic annotation. They transferred the things in IoT-System to a smart object by describing their functional and non-functional properties. The proposed framework contains four stages:

1. Virtual sensor layer (software for aggregating abstract data from different sensors and connect them by using open-source project: Global Sensor Networks (GSN)).
2. Service layer creates web services by using REST and SOAP web services for representing different sensors and access data from them.
3. Semantic layer is the semantically annotating layer these services by using lightweight semantic language "SA-WSDL" based on OWL ontology.
4. Interoperability layer is the integration level and enables the combination of described different-services to execute more complex task.

Unfortunately, previous research improved the interoperability issue without any confidentiality or security. IoT-systems that improved interoperability requires to be more secure with trustworthy, privacy and confidentiality between its members. Furthermore, healthcare systems take interoperability as a crucial issue.

By talking about interoperability as a basic issue [33, 52, 53], in Fig. 13 we introduce an IoT-Environment with requirements to achieve it on IoT-platform (using ontology methods, Interoperability Issue, Security mechanisms). The new ecosystem will provide security and confidentiality of things/objects with interoperablity. Based-ontology technique is implemented to achieve the interoperability and security over IoT-Context.

On the other hand, Huang et al. [54] created an ontology trust model to specify the semantics of trust in clear vision and support making trust judgment based on social network on the web. Due to the fact that not all trust are transitive, they identified the *semantics of trust*. By which a logical model of trust is created to expose the logical relations on the constructs of trust. To prove the transitivity: $(\forall x)(holds(trust\ b(d, c, x, k), s)) \wedge (\forall x)(holds(trust\ p(c, e, x, q), s)) \supset (\forall x)(holds(trust\ p(d, e, x, k \wedge q), s))$.

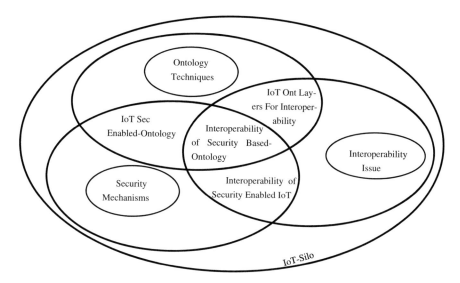

Fig. 13 Internet of things context

Unfortunately, the authors proved only two from the three parts in the implications of trust. Furthermore, their certainty model was not effective through cyberspaces. Contrariwise, trust model based accurate Trust Value (TV) materialized using an innovative Adaptable Cloud Inference System (ACIS) by Athanasiou [55]. TV estimation is performed via fuzzy-probabilistic reasoning that deals with uncertainty, hidden nonlinearities, and facility for proficient discrimination between trustworthy and untrustworthy. Ubiquitous Healthcare (UH) provides inside clean environment under malicious attacks.

New vision was provided by [56] into trust development by reconciling the two perspectives (Interpersonal, and organizational) trust. They based their theory on the fact that purchase intention relies on four boundary conditions. Also, they augmented the regression analysis with Fuzzy-set Qualitative Comparative Analysis (fsQCA). Osman et al. [57] proposed a "trust model that ascertains the expectation regarding agent's future presentation in a given setting by evaluating both the agent's readiness and ability through the semantic correlation of the current setting being referred to with the agent's exhibition in past comparable encounters". Their model depends on the previous experiences to predict if the agent is both qualified and ready to carry out the action that is designed to perform. To achieve IoT integration, aspects like feelings integrity and interoperability was not taken in consideration by the authors.

Bhattacharya in [58] argued that the agency of human related with care services must transfer to material agency of IoT-enabled smart technology depending on authentication value and trust to design such services. They presented a framework with three stages of trust to exploit the value co-creation. They achieved enhancing the trustworthy and authentication among tele-medical care advances and future plans of related help plans of action business models.

The interactions through objects may cause the creation of augmented relationship graphs, like those present in online social networks. This poses a direct threat to the privacy as an avid attacker may not only know about the objects but also with whom they interact. Privacy preserving is another security challenge to be achieved, which emerged from integrating sensors in the internet. Deployment of autonomous objects in IoT that sense people's private information like patient sensitive data, pose a new scale of threat to individuals' privacy.

In contrast to conventional scenarios that users should take some tasks like keyword searching or data posting to put their privacy at risk, IoT nodes gathers people's private data without even noticing them [59]. Henze in [60] presented cloud-based privacy enforcement approach that is driven by users in IoT that concentrates on privacy preserving for end users. The authors in [61] proposed a Privacy preserving Aggregation protocol "PAgIoT" that is appropriate for IoT settings and enables multi-attribute gathering for sets of entities while allowing correlation of privacy-preserving value. A lightweight privacy-preserving trust model had been designed for reducing privacy leakage in the existence of untrusted service providers, so that providers can be prohibited from showing information to other parties for harmful uses [62]. Also, Conditional Privacy-reserving authentication with Access Link ability (CPAL) for roaming service is proposed to supply universal secure roaming service and multilevel privacy preservation [53].

Finally, we need to reach and satisfy high level of security over IoT- environment through IoT layers and achieve semantic interoperability level over the members in IoT-siloes to be more scalable. For example, the heterogeneity of Electronic Health Records (EHRs) in health-IoT systems provided from healthcare data. These sensitive data collected in daily clinical practice among numerous service providers. Complexity in accessing and reusing such data is seen as a pivotal challenge. Hence, we need comprehensive data models and a layer of semantic interoperability mechanisms as in [63] through the healthcare systems to enable data reuse and achieve integrity. Relying on unified structural/terminological interoperability framework based on Basic Formal Ontology mechanism. This framework could have been more attractive if it had some trustworthy and privacy-preserving between its members as the combination occurred in [64]. The researchers combined between privacy and trust in Social IoT (SIoT) environment, Trust ensures trust between the social objects, while privacy secures the critical personal information. They proposed an architecture for privacy rules preservation and trust maintenance in SIoT based on edge-crowd integration. For the implementation of the proposed solution, small-edge servers are used as crowd sources.

7 Healthcare-IoT (IoHT) Challenges

In healthcare based IoT, one of the most pivotal topics is the data security and privacy that captured and transmitted in real time to the network. Also, the complete integration between patients and service providers (clinician, nurses and hospitals) through the heterogeneous data make it trustworthy between members to reach a cooperative IoT healthcare system. Users may still face some challenges while handling security tokens across heterogeneous platforms and trust domains. Semantic gaps and unsuitability are main boundaries for trust information exchange in federated trust management.

Understanding IoT vision is not a simple job due to a lot of challenges that requires to be treated. Examples of main challenges includes availability, reliability, performance, scalability, interoperability, security, management, privacy preserving and trust. Facing these challenges enables service providers and application developers to develop their services efficiently. For example, security and privacy play an important role in all markets globally regarding to the sensitivity of consumers' privacy. Besides that, evaluating IoT services performance is a main challenge [6, 8, 42, 50, 65]. Although the great benefits of IoT integration in healthcare, there are many challenges regarding data storing, data management and data transferring between devices, security, privacy, unified and ubiquitous access.

Healthcare sensitive data is a profitable target for attackers. That's why, the primary motivation of healthcare providers is securing protected health information. [26, 30]. Otherwise, the use of the tiny sensors in health care impose a number of challenges due to restrictions in network capacity, computer specifications, memory limits and limited power supply for which various technologies need to be integrated and managed [28, 42, 44]. The use of numerous IoT technologies, as required in healthcare, generates a heterogeneous ecosystem that complicates the system and lowers its applicability. The interoperability issue becomes a significant area in IoT systems in general and in medical systems in particular [28, 45, 48, 66, 84].

In addition to the present challenge of developing an integrated heterogeneous health-IoT system, being able to maintain privacy and trustworthy of the data is another significant concern. The Healthcare industry requires security and trustworthy in the data manipulation process due to patients' medical data confidentiality and protection legislations. Recent days, sharing of records and information turns out to be increasingly predominant on cloud storage and mobile devices due to its availability anywhere anytime. Although, data on the cloud increases the danger of malicious attacks and the disclosure of private data on shared storage [44, 66–70, 85].

There are various medical image encryption protocols that have been presented based on chaotic systems in order to produce encryption keys, although there is no unanimous formal definition of a chaotic system [86]. Quantum walks are mathematical models of quantum physical systems that are very sensitive to initial conditions; moreover, extracting data out of quantum walks is a probabilistic process described as a nonlinear behavior [87]. Furthermore, quantum walks have properties like stability, non-periodicity, and have unlimited key space theoretically to withstand diverse attacks. Consequently, we may think of quantum walks as a chaotic system thus suitable to be used in cryptography [86–89].

Below is a list of challenges in health-IoT environment:

- Security and privacy challenges in network of Health facilities and hospitals.

- Lack of interoperability, due to the massive number of medical service providers and medical device manufactures.
- Difficulty of managing the workflow intelligently, due to the massive amount of health information that are generated continuously.
- Lack of protocols and standards for interconnectivity and communication.
- Ensuring the confidentiality and the integrity of privacy patients' data in IoT based Multi-cloud in case of compromising through insider attack.
- Enhancing encryption and access control procedures to reach a satisfaction level of authentication and security.

8 IoT-Healthcare and Blockchain Integration

Health data is more sensitive and private. Consequently, sharing and exchange these data may raise the risk of exposure. To handle and organize medical big data to achieve a satisfied level of security in order to enhance treatment process, healthcare environment must be compatible with IoT-context based Blockchain technology. Such environment will provide the robustness against failure and data exposure [35, 71].

8.1 Blockchain (BC) Technology

The Blockchain Technology serves as a decentralized method to manage IoT framework tackling a large number of the above issues. In an IoT framework, the BC can keep an unchanging or immutable record of the historical knowledge of intelligent devices.

This element empowers the independence (autonomous) of smart devices without the requirement for a centralized authority. Therefore, Blockchain opens the portal to a movement of IoT circumstances that were strikingly troublesome, or even difficult to actualize [72, 73]. Also, one of the most invigorating limitation of Blockchain is the ability to keep up a decentralized, believed record of all transactions occurring in a framework. This capacity is major to enable the distinct compliances and regulatory essentials of Industrial IoT (IIoT) applications with no needs for relying upon a centralized model. Embracing an institutionalized shared correspondence (Peer-to-Peer P2P) model to process several millions of transactions between devices will essentially diminish the expenses related with installing, preserving and keeping up huge concentrated data centers. Moreover, this will disseminate calculation and storage needs over the millions of devices that structure IoT networks [42, 44]. Our daily decisions in our real world are taken based on relational calculation and trust with another entity to take trustworthy decision. Hence, the IoT-system that improved the interoperability need to be more secured and with confidentiality and trustworthy between its members. The decentralized, independent, and trustless capacities of the Blockchain make it an ideal section to transform into an essential segment of

IoT-context. BC is essentially a decentralized, appropriated, shared, and immutable database record that stores library of benefits. BC exchanges over a distributed P2P arrangement providing a global dispersed trust.

Blockchain systems are Distributed Ledgers Technology (DLT) with a mechanism to allow transactions to verify the transactions by a group of (consensus) unreliable actors called miners [42, 74]. It gives distributed, transparent, immutable, reliable and auditable records. The Blockchain can be concealed transparently and completely, enabling access to all data exchanges or transactions that have happened since the main transaction of the system. The miners are members liable for generating the Proof of Work (PoW) or mining. The idea of verification of work is one of the fundamentals to empower trustless agreement in Blockchain. The confirmation of work comprises of a computationally escalated job which is fundamental for blocks generation over the chain. It is intricate to understand and simultaneously effectively unquestionable once finished [42, 44]. When a miner finishes the PoW, it distributes the new block over the network and the remainder of this network checks its legitimacy before adding it to the chain. Beside trustless consensus, cryptography and shared ledger Blockchain present another technology called Smart Contract (SC) as shown in Fig. 14. The SC is computer program or protocol in which allows a contract to be executed automatically or enforced taking in consideration a collection of predefined conditions.

Blockchain can be a key empowering innovation to produce feasible security answers for testing IoT security issues [75–78]. In addition, the frameworks cooperate with one another, it's basic to have a concurrent interoperability standard, which is protected and valid. Without a strong bottom-top structure we will exert higher risks and attacks with each device added to the IoT-environment. What we need is a safe and reliable IoT network with security ensured and high level of trust and privacy preserved without being isolated within the same ecosystem. That is an extreme

Fig. 14 Blockchain technology factors

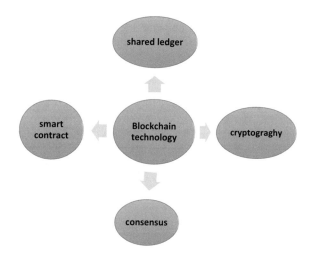

exchange off yet not feasible and BC innovation is an appealing alternative to achieve these needs [26, 27, 45].

8.2   Secure Health-IoT Models Based Blockchain

Blockchain engineering has a plenty of features that can be used in the healthcare environment as shown in Fig. 15.

These preferences are characteristic to the framework based Blockchain innovation and can be applied to a wide zone of frameworks and ventures. The highlights to be presented explicitly in this part are trustworthy, confirmation, integration and decentralized stockpiling. Blockchain technology in the healthcare context plays a key role in achieving security, trust and authentication between all members (doctors, hospital, patients and other medical parts) in healthcare domain. Mettler et al. [76] described examples that showed the importance of Blockchain and how it offers distinct opportunities for employment in the healthcare sector. BC increased safety in relation to drugs, and reduced health-related follow-up costs. Identified benefits of BC compared to conventional distributed databases for biomedical/ health care applications and will robustly impact the balance of power among existing market players in healthcare [79].

A prototype of Blockchain technology that enhances and guarantees privacy, security, and accessibility for overseeing and sharing EMR information for malignant growth (cancer) patient care was proposed in [69]. In addition, BC accomplished fine-grained EMR information for authority and lessen the expense. This structure can be increasingly effective by expanding the structure of a patient record and its metadata, utilizing the semantics of health information. The capability to robustly and securely construct privacy-preserving predictive model on healthcare data is an essential goal to be achieved [80]. Authors integrated online machine learning with Blockchains by constructing a system (Model Chain) to reach high level of

Fig. 15 The characteristics of blockchain

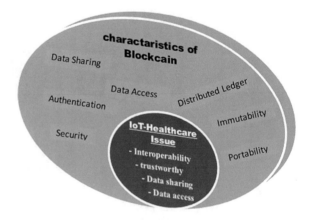

confidentiality and accomplished privacy-protection. Research can be done without uncovering any patient sensitive health data. Moreover, they structured algorithms for evidence-based decision-making working in real-time systems and online which need higher interoperability between various infrastructures. Healthcare industry is totally based on data that are gathered from patient or any person through health system. These healthcare data often arise some concerns over data confidentiality and privacy. Authors in [81, 82] suffered from two problems: first, they had considered storing and sharing only EMRs and ignored the beneficial and plentiful Personal Healthcare Data (PHD). The requirements of storing and sharing the enormous measure of PHD was viewed as more fundamentally from putting away and sharing the EMRs. Which gave new difficulties in situations like throughput and honorableness. Second, the current frameworks stored the EHRs in the cloud climate with sophisticated access control approach to block undesired information dissemination [70]. However, this framework vigorously depends on the security of the cloud space. Others studied more advanced systems to acknowledge persistent examination of healthcare information and advance the examination to reach precision medicine [83]. The authors provided an inclusive roadmap guide on how to implement such systems. Their framework or application should be implemented and experimented to assess their qualities and shortcomings.

Several of the proposed structures, frameworks, models and designs, for example [83], should be actualized and tried to assess their characteristics and shortcomings. Besides, for completely received Blockchain in operational healthcare conditions the interoperability open principles should be addressed to ensure communication between diverse Blockchain products. It is significant that analysts start investigating the trustworthy and interoperability issues and the institutionalization forms. The challenges of data security and authentication, interoperability, integrity and speed that depict BC-based healthcare applications are largely open research. Such research requires purposeful commitment so as to improve partners' trust in the utilization of the innovation and to advance its reception in medicinal services [71, 82, 96].

9 IoT-Healthcare Based Edge Computing

The Edge computing paradigm establishes another idea in the computing scene [90]. It fetches the utilities and services of distributed computing nearer to the end client and is portrayed by quick processing and speedy application response time [91, 92]. End clients typically run these applications on their asset compelled cell phones while the center assistance and handling are performed on cloud servers. Edge paradigm satisfies the previously mentioned application necessities by carrying preparations to the edge of the system. Among the promising highlights of Edge processing incorporated with portability support, location sensibility, ultra-low reaction time, and vicinity to the client [92–94]. These highlights make Edge processing appropriate for distinct future applications like smart home, smart health system, industrial systems, and traffic monitoring and smart meters [90, 95].

The security and privacy became a big challenge in cloud computing to guarantee satisfied level of security and trustworthy for end user, through low computational power [92, 93]. Differential-privacy strategy over edge paradigm achieved the above challenges [90, 91, 94].

10 Conclusion

The internet of things influence become prevalent and dominant, variety of smart devices and sensors use the internet to gather data. Traditional devices become clever and autonomous and/or independent. Due to advances in technology, smart devices are capable of communicating and trading information between each other. Combination between IoT and healthcare environments create a comfortable atmosphere to both patients and doctors/providers for effective illness monitoring and diagnosis. However, there are some challenges facing today's healthcare communication, particularly in the security domain e.g. (trustworthy and data privacy when interoperability is instigated). In this chapter, we introduced an overview about what is IoT environment with its vision and components. The layers of IoT architecture is explained. Also, this chapter presented IoT challenges and application especially in healthcare systems and the role of the IoT-context in developing it. Furthermore, we presented IoHT benefits and issues like interoperability and security and presented a literature review for these issues. This chapter covered the limitations of healthcare systems in terms of lack of privacy or trustworthy between members over IoT-networks. Then, an introduction of Blockchain (BC) technology was given and its role in IoHT systems. Finally, we investigated state of the art researches that used security models based trustworthy or privacy protocols for enhanced treatment processes.

References

1. Singh, D., Tripathi, G., Jara, A.J.: A survey of internet-of-things: future vision, architecture, challenges and services. In: 2014 IEEE world forum on Internet of Things (WF-IoT), pp. 287–292. IEEE, (2014)
2. Reyna, A., Martín, C., Chen, J., Soler, E., Díaz, M.: On blockchain and its integration with IoT. Challenges and opportunities. Future Gener. Comput. Syst. **88**, 173–190 (2018)
3. Tarouco, L.M.R., Bertholdo, L.M., Granville, L.Z., Arbiza, L.M.R., Carbone, F., Marotta, M., De Santanna, J.J.C.: Internet of things in healthcare: interoperatibility and security issues. In: 2012 IEEE International Conference on Communications (ICC), pp. 6121–6125. IEEE (2012)
4. Palattella, M.R., Accettura, N., Grieco, L.A., Boggia, G., Dohler, M., Engel, T.: On optimal scheduling in duty-cycled industrial IoT applications using IEEE802. 15.4 e TSCH. IEEE Sens. J. **13**(10), 3655–3666 (2013)
5. Wang, X., Zha, X., Ni, W., Liu, R.P., Guo, Y.J., Niu, X., Zheng, K.: Survey on blockchain for internet of things. Comput. Commun. **136**, 10–29 (2019)
6. McGhin, T., Choo, K.-K.R., Liu, C.Z., He, D.: Blockchain in healthcare applications: Research challenges and opportunities. J. Netw. Comput. Appl. **135**, 62–75 (2019)

7. Zhang, M., Yu, T., Zhai, G.F:. Smart transport system based on the internet of things. In: Applied Mechanics and Materials, vol. 48, pp. 1073–1076. Trans Tech Publications Ltd, (2011)

8. Yun, M., Yuxin, B.: Research on the architecture and key technology of internet of things (IoT) applied on smart grid. In: 2010 International Conference on Advances in Energy Engineering, pp. 69–72. IEEE (2010)

9. Al-Fuqaha, A., Guizani, M., Mohammadi, M., Aledhari, M., Ayyash, M.: Internet of things: a survey on enabling technologies, protocols, and applications. IEEE Commun. Surv. Tutor.17(4), 2347–2376 (2015)

10. Compton, M., Henson, C.A., Lefort, L., Neuhaus, H., Sheth, A.P.: A survey of the semantic specification of sensors, p. 17 (2009)

11. Cerullo, G., Mazzeo, G., Papale, G., Ragucci, B., Sgaglione, L.: IoT and sensor networks security. In: Security and Resilience in Intelligent Data-Centric Systems and Communication Networks, pp. 77–101. Academic Press (2018)

12. O'Conner, L.: 2016 IEEE tenth international conference on semantic computing, ICSC 2016. Laguna Hills, California: proceedings. In: 10th IEEE International Conference on Semantic Computing. IEEE Computer Society (2016)

13. Serrano, M., Barnaghi, P., Carrez, F., Cousin, P., Vermesan, O., Friess, P.: Internet of things IoT semantic interoperability: research challenges, best practices, recommendations and next steps. IERC: Eur. Res. Cluster Internet Things, Tech. Rep. (2015)

14. Yang, Z., Yue, Y., Yang, Y., Peng, Y., Wang, X., Liu, W.: Study and application on the architecture and key technologies for IOT. In: 2011 International Conference on Multimedia Technology, pp. 747–751. IEEE (2011)

15. [15] Al-Fuqaha, A., Guizani, M., Mohammadi, M., Aledhari, M., Ayyashm M.: Internet of things: a survey on enabling technologies, protocols, and applications. IEEE Commun. Surv. Tutor. 17(4), 2347–2376 (IEEE) (2015)

16. Koshizuka, N., Sakamura, K.: Ubiquitous ID: standards for ubiquitous computing and the internet of things. IEEE Pervasive Comput. 9(4), 98–101 (2010)

17. Ferro, E., Potorti, F.: Bluetooth and Wi-Fi wireless protocols: a survey and a comparison. IEEE Wirel. Commun. 12(1), 12–26 (2005)

18. McDermott-Wells, P.: What is bluetooth? IEEE Potentials 23(5), 33–35 (2005)

19. Press releases detail: bluetooth technology website. Bluetooth Technol. Website, Kirkland, WA, USA (2014)

20. IEEE Standard for Local and Metropolitan Area Networks—Part 15.4: Low-Rate Wireless Personal Area Networks (LR-WPANs), IEEE Std. 802. 15, 4–2011 (2011)

21. Barnaghi, P., Wang, W., Henson, C., Taylor, K.: Semantics for the internet of things: early progress and back to the future. Int. J. Semant. Web Inf. Syst. (IJSWIS) 8(1), 1–21 (2012)

22. Desai, P., Sheth, A., Anantharam, P.: Semantic gateway as a service architecture for IoT inter-operability. In: 2015 IEEE International Conference on Mobile Services, pp. 313–319. IEEE (2015)

23. Huang, Y., Li, G.: A semantic analysis for internet of things. In: 2010 International Conference on Intelligent Computation Technology and Automation, vol. 1, pp. 336–339. IEEE (2010)

24. Schneider, J., Kamiya, T., Daniel Peintner, and Rumen Kyusakov. "Efficient XML interchange (EXI) format 1.0." W3C Proposed Recommendation 20, pp. 32, 2011.

25. Atzori, L., Iera, A., Morabito, G.: The internet of things: a survey. Comput. Netw. 54(15), 2787–2805 (2010)

26. Want, R.: Near field communication. IEEE Pervasive Comput. 10(3), 4–7 (2011)

27. Dang, L.M., Piran, Md., Han, D., Min, K., Moon, H.: A survey on internet of things and cloud computing for healthcare. Electronics 8(7), 768 (2019)

28. Kelati, A., Dhaou, I.B., Tenhunen, H.: Biosignal monitoring platform using wearable IoT. In: Proceedings of the 22st Conference of Open Innovations Association FRUCT, Petrozavodsk, Russia, pp. 9–13 (2018)

29. Zgheib, R., Conchon, E., Bastide, R.: Engineering IoT healthcare applications: towards a semantic data driven sustainable architecture. In: eHealth 360°, pp. 407–418. Springer, Cham (2017)

30. Xiao, G., Guo, J., Da Xu, L., Gong, Z.: User interoperability with heterogeneous IoT devices through transformation. IEEE Trans. Ind. Informat. **10**(2), 1486–1496 (2014)
31. Hussain, Md.I.: Internet of things: challenges and research opportunities. CSI Trans. ICT **5**(1), 87-95 (2017)
32. Gyrard, A., Serrano, M., Atemezing, G.A.: Semantic web methodologies, best practices and ontology engineering applied to internet of things. In: 2015 IEEE 2nd World Forum on Internet of Things (WF-IoT), pp. 412–417. IEEE (2015)
33. Ganzha, M., Paprzycki, M., Pawłowski, W., Szmeja, P., Wasielewska, K.: Semantic interoperability in the internet of things: an overview from the INTER-IoT perspective. J. Netw. Comput. Appl. **81**, 111–124 (2017)
34. Strassner, J., Diab, W.W.: A semantic interoperability architecture for internet of things data sharing and computing. In: 2016 IEEE 3rd World Forum on Internet of Things (WF-IoT), pp. 609–614. IEEE (2016)
35. Kshetrimayum, R.S.: An introduction to UWB communication systems. IEEE Potentials **28**(2), 9–13 (2009)
36. Pramanik, P.K.D., Pareek, G., Nayyar, A.: Security and privacy in remote healthcare: issues, solutions, and standards. In: Telemedicine Technologies, pp. 201–225. Academic Press (2019)
37. Catarinucci, L., De Donno, D., Mainetti, L., Palano, L., Patrono, L., Stefanizzi, M.L., Tarricone, L.: An IoT-aware architecture for smart healthcare systems. IEEE Internet Things J. **2**(6), 515–526 (2015)
38. Ray, P.P., Dash, D., De, D.: Edge computing for internet of things: a survey, e-healthcare case study and future direction. J. Netw. Comput. Appl. **140**, 1–22 (2019)
39. Pang, Z.: Technologies and architectures of the internet-of-things (IoT) for health and well-being. Ph.D. diss., KTH Royal Institute of Technology (2013)
40. Islam, S.M.R., Kwak D., Kabir, M.D.H., Hossain, M., Kwak, K.-S.: The internet of things for health care: a comprehensive survey. IEEE Access **3**, 678–708 (2015)
41. Dwivedi, A.D., Srivastava, G., Dhar, S., Singh, R.: A decentralized privacy-preserving healthcare blockchain for IoT. Sensors **19**(2), 326 (2019)
42. Erdeniz, S.P., Maglogiannis, I., Menychtas, A., Felfernig, A., Tran, T.N.T.: Recommender systems for IoT enabled m-health applications. In: IFIP International Conference on Artificial Intelligence Applications and Innovations, pp. 227–237. Springer, Cham (2018)
43. Athanasiou, G., Anastassopoulos, G.C., Tiritidou, E., Lymberopoulos, D.: A trust model for ubiquitous healthcare environment on the basis of adaptable fuzzy-probabilistic inference system. IEEE J. Biomed. Health Inform. **22**(4), 1288–1298 (2017)
44. AlMotiri, S.H., Khan, M.A., Alghamdi, M.A.: Mobile health (m-health) system in the context of IoT. In: 2016 IEEE 4th International Conference on Future Internet of Things and Cloud Workshops (FiCloudW), pp. 39–42. IEEE (2016)
45. Jabbar, S., Ullah, F., Khalid, S., Khan, M., Han, K.: Semantic interoperability in heterogeneous IoT infrastructure for healthcare. Wirel. Commun. Mob. Comput. 2017 (2017)
46. Ha, M., Kim, S.H., Kim, D.: Intra-mario: a fast mobility management protocol for 6lowpan. IEEE Trans. Mob. Comput. **16**(1), 172–184 (2016)
47. Reda, R., Piccinini, F., Carbonaro, A.: Semantic modelling of smart healthcare data. In: Proceedings of SAI Intelligent Systems Conference, pp. 399–411. Springer, Cham (2018)
48. Ullah, F., Habib, M.A., Farhan, M., Khalid, S., Durrani, M.Y., Jabbar, S.: Semantic interoperability for big-data in heterogeneous IoT infrastructure for healthcare. Sustain. Cities Soc. **34**, 90–96 (2017)
49. Want, R.: An introduction to RFID technology. IEEE Pervasive Comput. **5**(1), 25–33 (2006)
50. Singh, S., Singh, N.: Internet of things (IoT): security challenges, business opportunities & reference architecture for E-commerce. In: 2015 International Conference on Green Computing and Internet of Things (ICGCIoT), pp. 1577–1581. IEEE (2015)
51. Blackstock, M., Lea, R.: IoT interoperability: a hub-based approach. In: 2014 International Conference on the Internet of Things (IOT), pp. 79–84. IEEE (2014)
52. Serrano, M., Quoc, H.N.M., Phuoc, D.L., Hauswirth, M., Soldatos, J., Kefalakis, N., Jayaraman, P.P., Zaslavsky, A.: Defining the stack for service delivery models and interoperability in the

internet of things: a practical case with Open IoT-VDK. IEEE J. Sel. Areas Commun. **33**(4), 676–689 (2015)

53. Androcec, D., Vrcek, N.: Thing as a service interoperability: review and framework proposal. In: 2016 IEEE 4th International Conference on Future Internet of Things and Cloud (FiCloud), pp. 309–316. IEEE (2016)

54. Lai, C., Li, H., Liang, X., Rongxing, L., Zhang, K., Shen, X.: CPAL: a conditional privacy-preserving authentication with access linkability for roaming service. IEEE Internet Things J. **1**(1), 46–57 (2014)

55. Huang, J., Fox, M.S.: An ontology of trust: formal semantics and transitivity. In: Proceedings of the 8th International Conference on Electronic Commerce: the New e-Commerce: innovations for Conquering Current Barriers, Obstacles and Limitations to Conducting Successful Business on the Internet, pp. 259–270 (2006)

56. Li, L., Xiaoguang, H., Ke, C., Ketai, H.: The applications of wifi-based wireless sensor network in internet of things and smart grid. In: 2011 6th IEEE Conference on Industrial Electronics and Applications, pp. 789–793. IEEE (2011)

57. Zheng, S., Hui, S.F., Yang, Z.: Hospital trust or doctor trust? A fuzzy analysis of trust in the health care setting. J. Bus. Res. **78**, 217–225 (2017)

58. Osman, N., Sierra, C., Mcneill, F., Pane, J., Debenham, J.: Trust and matching algorithms for selecting suitable agents. ACM Trans. Intell. Syst. Technol. (TIST) **5**(1), 1–39 (2014)

59. Bhattacharya, S., Wainwright, D., Whalley, J.: internet of things (IoT) enabled assistive care services: designing for value and trust. Procedia Comput. Sci. **113**, 659–664 (2017)

60. Lopez, J., Rios, R., Bao, F., Wang, G.: Evolving privacy: from sensors to the internet of things. Future Gener. Comput. Syst. **75**, 46–57 (2017)

61. Henze, M., Hermerschmidt, L., Kerpen, D., Häußling, R., Rumpe, B., Wehrle, K.: A comprehensive approach to privacy in the cloud-based internet of things. Future Gener. Comput. Syst. **56**, 701–718 (2016)

62. González-Manzano, L., de Fuentes, J.M., Pastrana, S., Peris-Lopez, P., Hernández-Encinas, L.: PAgIoT—Privacy-preserving aggregation protocol for internet of things. J. Netw. Comput. Appl. **71**, 59–71 (2016)

63. Appavoo, P., Chan, M.C., Bhojan, A., Chang, E.-C.: Efficient and privacy-preserving access to sensor data for internet of things (IoT) based services. In: 2016 8th International Conference on Communication Systems and Networks (COMSNETS), pp. 1–8. IEEE (2016)

64. Ethier, J.F., Curcin, V., Barton, A., McGilchrist, M.M., Bastiaens, H., Andreasson, A., Delaney, B.C.: Clinical data integration model. Methods Inf. Med. **54**(01), 16–23 (2015)

65. Sharma, V., You, I., Jayakody, D.N.K., Atiquzzaman, M.: Cooperative trust relaying and privacy preservation via edge-crowdsourcing in social internet of things. Future Gener. Comput. Syst. **92**, 758–776 (2019)

66. Conti, M., Dehghantanha, A., Franke, K., Watson, S.: Internet of Things Security and Forensics: challenges and Opportunities, pp. 544–546 (2018)

67. Yang, J.-J., Li, J.-Q., Niu, Y.: A hybrid solution for privacy preserving medical data sharing in the cloud environment. Future Gener. Comput. Syst. **43**, 74–86 (2015)

68. Yang, Y., Zheng, X., Tang, C.: Lightweight distributed secure data management system for health internet of things. J. Netw. Comput. Appl. **89**(2017), 26–37 (2017)

69. Curcin, V., Bastiaens, H., Taweel, A.: Clinical data integration model: core interoperability ontology for research using primary care data. Methods Inf. Med. **54**, 16–23 (2015)

70. Dubovitskaya, A., Xu, Z., Ryu, S., Schumacher, M., Wang, F.: Secure and trustable electronic medical records sharing using blockchain. In: AMIA Annual Symposium Proceedings, vol. 2017, p. 650. American Medical Informatics Association (2017)

71. Al Omar, A., Bhuiyan, Md.Z.A., Basu, A., Kiyomoto, S., Rahman, M.S.: Privacy-friendly platform for healthcare data in cloud based on blockchain environment. Future Gener. Comput. Syst. **95**, 511–521 (2019)

72. Griggs, K.N., Ossipova, O., Kohlios, C.P., Baccarini, A.N., Howson, E.A., Hayajneh, T.: Healthcare blockchain system using smart contracts for secure automated remote patient monitoring. J. Med. Syst. **42**(7), 130 (2018)

73. Aggarwal, S., Chaudhary, R., Aujla, G.S., Kumar, N., Choo, K.-K.R., Zomaya, A.Y.: Blockchain for smart communities: applications, challenges and opportunities. J. Netw. Comput. Appl. **144**, 13–48 (2019)
74. Moin, S., Karim, A., Safdar, Z., Safdar, K., Ahmed, E., Imran, M.: Securing IoTs in distributed blockchain: analysis, requirements and open issues. Future Gener. Comput. Syst. **100**, 325–343 (2019)
75. Lu, Y.: The blockchain: state-of-the-art and research challenges. J. Ind. Inf. Integr. **15**, 80–90 (2019)
76. Abdullah, N., Hakansson, A., Moradian, E.: Blockchain based approach to enhance big data authentication in distributed environment. In: 2017 Ninth International Conference on Ubiquitous and Future Networks (ICUFN), pp. 887–892. IEEE (2017)
77. Mettler, M.: Blockchain technology in healthcare: the revolution starts here. In: 2016 IEEE 18th International Conference on e-Health Networking, Applications and Services (Healthcom), pp. 1–3. IEEE (2016)
78. Khan, M.A., Salah, K.: IoT security: review, blockchain solutions, and open challenges. Future Gener. Comput. Syst. **82**, 395–411 (2018)
79. Banerjee, M., Lee, J., Choo, K.-K.R.: A blockchain future for internet of things security: a position paper. Digital Commun. Netw. **4**(3), 149–160 (2018)
80. Kuo, T.-T., Kim, H.-E., Ohno-Machado, L.: Blockchain distributed ledger technologies for biomedical and health care applications. J. Am. Med. Inform. Assoc. **24**(6), 1211–1220 (2017)
81. Kuo, T.-T., Ohno-Machado, L.: Modelchain: decentralized privacy-preserving healthcare predictive modeling framework on private blockchain networks (2018). arXiv:1802.01746
82. Shae, Z., Tsai, J.P.: On the design of a blockchain platform for clinical trial and precision medicine. In: 2017 IEEE 37th International Conference on Distributed Computing Systems (ICDCS), pp. 1972–1980. IEEE (2017)
83. Chen, L., Lee, W.-K., Chang, C.-C., Choo, K.-K.R., Zhang, N.: Blockchain based searchable encryption for electronic health record sharing. Future Gener. Comput. Syst. **95**, 420–429 (2019)
84. Mamoshina, P., Ojomoko, L., Yanovich, Y., Ostrovski, A., Botezatu, A., Prikhodko, P., Izumchenko, E., et al.: Converging blockchain and next-generation artificial intelligence technologies to decentralize and accelerate biomedical research and healthcare. Oncotarget **9**(5), 5665 (2018)
85. Abd El-Latif, A.A., Hossain, M.S., Wang, N.: Score level multibiometrics fusion approach for healthcare. Cluster Comput. **22**(1), 2425–2436 (2019)
86. Gad, R., Abd El-Latif, A.A., Elseuofi, S., Ibrahim, H.M., Elmezain, M., Said, W.: IoT security based on iris verification using multi-algorithm feature level fusion scheme. In: 2019 2nd International Conference on Computer Applications & Information Security (ICCAIS), pp. 1–6. IEEE (2019)
87. Tsafack, N., Sankar, S., Abd-El-Atty, B., Kengne, J., Jithin, K.C., Belazi, A., Mehmood, I., Bashir, A.K., Song, O.-Y., Abd El-Latif, A.A.: A new chaotic map with dynamic analysis and encryption application in internet of health things. IEEE Access **8**, 137731–137744 (2020)
88. Abd El-Latif, A.A., Abd-El-Atty, B., Venegas-Andraca, O.-Y., Elwahsh, H., Piran, Md.J., Bashir, A.K., Song, O.-Y., Mazurczyk, W.: Providing end-to-end security using quantum walks in IoT networks. IEEE Access (2020)
89. Abd-El-Atty, B., Iliyasu, A.M., Alaskar, H., El-Latif, A., Ahmed, A.: A robust quasi-quantum walks-based steganography protocol for secure transmission of images on cloud-based E-healthcare platforms. Sensors **20**(11), 3108 (2020)
90. Abd EL-Latif, A.A., Abd-El-Atty, B., Abou-Nassar, E.M., Venegas-Andraca, S.E.: Controlled alternate quantum walks based privacy preserving healthcare images in internet of things. Opt. Laser Technol. **124**, 105942 (2020)
91. Gheisari, M., Wang, G., Chen, S.: An Edge Computing-enhanced internet of things framework for privacy-preserving in smart city. Comput. Electr. Eng. **81**, 106504 (2020)
92. Wang, T., Mei, Y., Jia, W., Zheng, X., Wang, G., Xie, M.: Edge-based differential privacy computing for sensor—cloud systems. J. Parall. Distrib. Comput. **136**, v75-85 (2020)

93. Gad, R., Talha, M., Abd El-Latif, A.A., Zorkany, M., Ayman, E.-S., Nawal, E.-F., Muhammad, G.: Iris recognition using multi-algorithmic approaches for cognitive internet of things (ciot) framework. Future Gener. Comput. Syst. **89**, 178–191 (2018)
94. Zhang, W.-Z., Elgendy, I.A., Hammad, M., Iliyasu, A.M., Du, X., Guizani, M., Abd El-Latif, A.A.: Secure and optimized load balancing for multi-tier IoT and edge-cloud computing systems. IEEE Internet Things J. (2020)
95. Abou-Nassar, E.M., Iliyasu, A.M., El-Kafrawy, P.M., Song, O.-Y., Bashir, A.K., Abd El-Latif, A.A.: DITrust chain: towards blockchain-based trust models for sustainable healthcare IoT systems. IEEE Access **8**, 111223–111238 (2020)
96. Alghamdi, A., Hammad, M., Ugail, H., Abdel-Raheem, A., Muhammad, K., Khalifa, H.S., Abd El-Latif, A.A.: Detection of myocardial infarction based on novel deep transfer learning methods for urban healthcare in smart cities. Multimed. Tools Appl. 1–22 (2020)
97. Masud, M., Gaba, G.S., Alqahtani, S., Muhammad, G., Gupta, B.B., Kumar, P., Ghoneim, A.: A lightweight and robust secure key establishment protocol for internet of medical things in COVID-19 patients care. IEEE Internet Things J. (2020)

Assisted Fog Computing Approach for Data Privacy Preservation in IoT-Based Healthcare

Mohamed Sarrab and Fatma Alshohoumi

Abstract Recently, the internet of things (IoT) technologies plays a very important role in various important sectors such as healthcare, education and industry. It has changed the conventional way of diagnosing some diseases and accelerating the check-up process through using IoT medical devices. Many IoT devices are available for measuring the biomarkers such as heart rate, sugar level and blood pressure, etc. However, the privacy of these data collected via IoT medical devices remains a challenge that hinders the use of these devices in clinical practice. The massive data collected by these devices vary in their sensitivity to patients. The more sensitive data requires fast computation and processing to avoid any delay that may occur. The processing of such data in the cloud may lead to operations delay which is needed by a real-time monitoring application. Therefore, this research intends to provide Healthcare Internet of Thing (H-IoT)-based framework for the classification of streamed data according to their criticality level. After the classification, the more crucial data will be computed in fog rather than in the cloud to avoid latency. Future work will extend this work by implementing the proposed framework and evaluating its outcomes.

Keywords Fog computing · Internet of things (IoT) · Data privacy · Privacy preservation · Healthcare internet of thing (H-IoT) · Internet of medical things

1 Introduction

IoT has attracted the attention of academia and industries in recent years. An IoT is focusing on facilitating the interaction between humans and surrounding things [1]. In the IoT environment, objects/things are generally used to gather and receive data, where the data computing is done remotely in the cloud [2]. H-IoT have a major role

M. Sarrab (✉) · F. Alshohoumi
Communication and Information Research Center, Sultan Qaboos University, Muscat, Oman
e-mail: sarrab@squ.edu.om

F. Alshohoumi
e-mail: alshohumi@squ.edu.om

© The Author(s), under exclusive license to Springer Nature Switzerland AG 2022
A. A. Abd El-Latif et al. (eds.), *Security and Privacy Preserving for IoT and 5G Networks*,
Studies in Big Data 95, https://doi.org/10.1007/978-3-030-85428-7_8

in leveraging medical and health care processes through improving several crucial IoT medical applications for observing patient's state. For example, hypertension attacks can be identified and controlled using IoT. In H-IoT, the risk of hypertension attack can be predicted using IoT-based medical sensors, which can provide users and doctors with real-time notifications to diagnose their health state [3].

The key functionality of H-IoT is to observe real-live information of the patients collected using IoT devices [4]. Although H-IoT has the potential for significant impact on health-related wellness and socio-economic growth, it comes across numerous obstacles associated with real-live observation e.g. response time, reliability, patients' privacy, and security of the IoT collected data [5, 6]. This work is designed to provide a fog-based approach focused on the classification of IoT collected data into several fog virtual servers based on their criticality or privacy level to help in preserving IoT data privacy and enhance the response time when noticing the irregular measurements.

2   Fog Computing

The idea of fog computing is concentrated on decentralized processing whereas; storage, data, applications and processing are located somewhere between the devices and the cloud. Similar to the concepts of edge processing, fog computing provides the power and advantages of the cloud together and brings them closer to the devices (data sources). Researchers are using the terms edge computing and fog computing interchangeably as both of them are focusing on bringing computing closer to the data source. The fog networking is not to replace cloud computing but to be as a complementary component to the decentralized processing infrastructure which takes place in-between the the cloud and IoT devices. In general, fog computing is used because the cloud technology is not practically feasible for different IoT applications.

Its decentralized approach considers the need for H-IoT, besides the big amount of data produced by IoT devices that can be time-consuming and costly to process and analyze in the cloud. Fog computing decreases the needed communication bandwidth and the back-and-forth communication between devices and the cloud, which will have a negative impact on the overall IoT ecosystem performance.

As show in Fig. 1, fog computing as a sort of decentralized processing infrastructure is located between the cloud (data storage) and edge devices. Fog computing improves the idea of cloud technology to the network edge, which makes it suitable for real-time information. Where; the ultimate goal of IoT technologies is real-time interactions and up-to-date information.

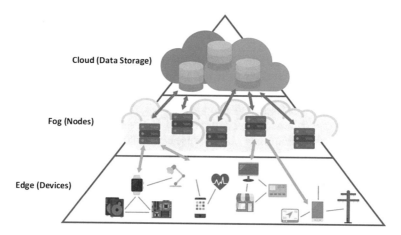

Fig. 1 Fog computing as a decentralized computing infrastructure

3   Privacy Preservation

IoT brings a high level of control and automation to our everyday tasks by connecting the surrounding environment with smart objects. IoT smart objects refer to a wide range of nonstandard computing devices including actuators, sensors and micro-controllers that are used to exchange data to support smarter interactions through machine to machine communications and enable informed decision making.

Considering the high penetration and impact of IoT technologies on our surrounding environment and everyday life activities, we need to realize the IoT technologies privacy associated challenges and risks. More importantly, we are in need to consider the IoT user privacy preservation and safe environment similar to those offered in traditional internet technologies.

It is very important to distinguish between different relevant terminologies such as security, trust and privacy. In general, security can be defined as the state of being free from threat or danger considering three main security principles confidentiality, integrity and availability. In IoT, security refers to the protection of devices and networks they are connected to from unauthorized access. However, trust is derived from two crucial concepts: consistency and transparency. Consistency refers to the behavior of IoT devices consistently to meet end users' expectations. Transparency refers to the data sources and how they inform end users about the purpose of the data gathering process, the sort of collected data, and how that data will be used. Whereas, privacy refers to the individual's information that must be protected and should not be exposed in the IoT environment without explicit individual's consent under any circumstances.

4   Related Work

IoT in the healthcare environment improves the accuracy in gathering vital signs in an errorless manner. IoT provides many great solutions to the healthcare system such as reducing the need for beds, doctors, and nurses. It also provides different smart and sustainable solutions for patients and the overall healthcare system [7].

Several IoT applications have been adopted in the healthcare field, including assistants for elderly patients and chronic disease trackers [8, 9] as well as, remote health monitoring (e.g., fall detection [10], echocardiograms (ECGs) [7], and glucose level monitoring [9], etc.). H-IoT has been used to facilitate the communication process between doctors and their patients [11]. Figure 2 depicts the general H-IoT architecture.

Figure 2 showed, the general H-IoT architecture composed of (1) H-IoT devices or sensors that are used for collecting the biomarkers from patients such as blood pressure, heartbeat, blood sugar, etc. (2) Gateways which aim to transfer the data collected by the sensors to cloud for the processing, (3) Data storage which can be performed in the edge nearby H-IoT devices or far in the cloud.

Several IoT medical devices have been developed to track the health status by providing different measurements such as heart rate, blood pressure, respiratory rate, etc. [12]. In H-IoT, the data gathered by sensors are transferred using different communication channels for further computing. The collected data can be used for different purposes where they can be evaluated to infer addictions, e.g. drinking or smoking [13]. The data collected by wrist-worn sensors are used to classify eating or smoking practices [14]. Moreover, the medical sensors are used to collect data to predict different types of diseases (e.g., stroke, heart attack, etc.), because given that, the precautionary process can be used for patient's protection considering any unexpected states.

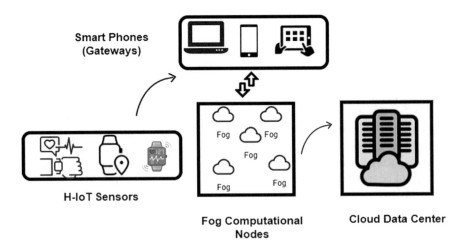

Fig. 2 The general H-IoT architecture

Even though H-IoT offers a variety of important approaches to different individuals including patients, doctors, nurses, and administrators, the collected data transfer from the sensors layer via the communication channel to the storage level is bounded by different security attacks and threats [15, 16]. e.g. IoT devices might incidentally expose critical important data (private information) such as smart switches or thermostats might be used to disseminate information about a home occupation at a specific time [17, 18]. Moreover, loss of control over data after sent for storage in the cloud, whereas the gathered data that received in cloud can be transferred to an untrusted third party[18].

In the last couple of years, the IoT healthcare system uses cloud computing to analyze and store the collected data [3]. Nevertheless, many issues and obstacles were faced by the cloud, mainly in H-IoT the use of cloud computing, surfaces resource-constrained devices, latency constraints, reliability and location awareness challenges [5]. Fog computing was introduced by Cisco [19] to overcome response time and network latency issues via its utilization of connectivity gateways, routers, and mobile base stations[20] [21]. Fog computing is a modern data processing model that has been provided to solve and overcome some cloud computing challenges. Fog computing supports the functions of lightweight networking services, processing, and storage of IoT collected data and the cloud. Furthermore, FC accomplishes IoT application requirements through the processing of collected data at a specific deadline. As reported by Islam et al., in 2015, the data collected by H-IoT devices can be categorized into small data (which can be processed in fog) and big data (that can be processed in the cloud) [9].

The healthcare data volume and velocity generated by H-IoT devices are massive, approximately 150 MB/min or more [20]. Processing these data closer to the H-IoT devices supports offering better services considering distributing risk notifications immediately and quicker response time if it compared with the cloud whereas, the delay might occur due to network congestion [2]. Therefore, fog computing is a suitable model for computing and processing data closer to data sources instead of processing and computing in the cloud [4, 22].

Many types of research were conducted focusing on IoT-fog-based systems. In 2018, Sandeep et al. designed a framework to control hypertension attacks based on IoT fog computing [3]. The fog was used for observing the patient's blood pressure and for real-time diagnose of the hypertension stage. Their proposed framework accomplished bandwidth efficiency, high accuracy, and low response time.

In 2019, Sukhpal et al. designed a new model for IoT-based fog-enabled cloud technology in healthcare field. Their approach is designed to control data of heart patients effectively. Where the data received from different IoT sensors to diagnose patients' medical status. The data of heart patients' will be processed using a fog-assisted cloud technology that was allocated near to the data sources to decrease response time, and save network bandwidth. As result, their proposed model showed that the approach of fog computing as compared to the cloud computing reduces the latency from 19.56 to 29.45% [20].

In 2019, Omar and Khalid, analyzed some of the fog-based IoT concerns to provide security scheme built on MQTT protocol [2]. Their discussion and analysis showed that MQTT protocol and security based on fog were appropriate to performance. Even though the mentioned research above used fog computing to offer low latency and save the bandwidth, however, the collected data that processed in fog devices are exposed to privacy and security concerns similar to the cloud. The research on data security (confidentiality, integrity and availability) and privacy in fog is still in its early stage [23].

In 2019, Elgendy et al. addressed the limitations of IoT devices such as computational power and the security of the sensitive information generated by multi-users. They presented a multiuser resource allocation and computation offloading model that is integrated with security to address the limitations. Advanced encryption standard cryptographic method was used as a security layer for protecting the sensitive data from any attacks. Moreover, the offloading algorithm was implemented for a purpose of determining the decision of offloading computation of the data generated by multi-users of IoT devices. The results of this study demonstrated that the model with or without the addition of the security layer can reduce or minimize the computation overhead in terms of time [24].

In 2020, blockchain technology has been integrated with an IoT-based healthcare framework for security and efficiency enhancement. The proposed model was based on decentralization and interoperability trust. It supported semantic annotations for edge layers in H-IoT. Cryptographic techniques were used across different stages of data inclusion and exchange for authentication, validation, and protection. The performance of this model was better than other approaches in terms of scalability, interoperability, availability, trustworthiness, data integrity, confidentiality, and data privacy [25].

In 2021, Elgendy et al. proposed a model for resource allocation data offloading integrated with security to assure efficient sharing of resources among multiple users of industrial IoT devices. The proposed model was based on an advanced deep reinforcement and aimed to minimize energy consumption and computational delay. The model was presented to find the best solution. Moreover, the 128-bit of advanced encryption standard was used as a security method for protecting the data. The experimental results revealed that the implemented model can reduce the time of the offloading overhead by 64.7%. According to the findings, this model can scale for a huge number of IoT-based mobile devices [26].

Generally, the streamed data send to fog are huge because they are collected from different IoT-based medical sensors. The collected data can be of different types ranging from health-related data to personally identifiable information. Accordingly, further data classification is needed to reduce the fog data size for getting more efficient and to preserve data privacy. Our proposed research attempts to utilize fog for data preprocessing. The data is going to be preprocessed by filtering the generated data based on their criticality level (non-private, private (critical), private (sensitive)). The useless data will be removed (cleaned) to accelerate data processing in fog. Using filtration or classification process will help in reducing the size of data and hence increasing the response time of processed data.

5 The Proposed Approach

The proposed approach focuses on collected data preservation in fog for IoT-based healthcare. The approach classifies the data received from different sensors based on their specified privacy level non-private (normal), private (critical), or high private (sensitive).

As shown in Fig. 3, the proposed approach consists of the following components.

5.1 IoT Medical Sensors

Various H-IoT sensors are developed to collect health biomarkers, for example, to measure patient blood pressure (BP), blood sugar, heartbeat (HR), respiration rate (RP), and temperature. H-IoT sensors can also send other attributes information such as personal information including name, phone, address, location, etc. The generated data will be sent to the edge (fog devices) and the cloud servers for preprocessing and analysis via wireless communication channels. This research proposed an approach to preserve and process the generated data in fog.

5.2 IoT Devices (Fog Nodes)

This component involved the hospital servers and specialized mobile phones which are dedicated to receiving the generated data by medical IoT devices for preprocessing. The purpose of handling the preprocessing of the generated data by IoT medical sensors is the location of fog nodes nearby to the H-IoT sensors which lead to effective real-time processing and decision processing. Such an approach offers

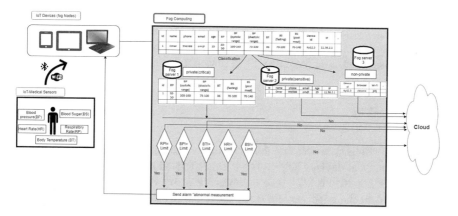

Fig. 3 Scenario of the proposed H-IoT approach of fog-based data privacy

minimum response time and latency for H-IoT applications that are needed to process health-related data within a specific deadline.

The following are four proposed main phases for fog-based data preprocessing:

1. Read data generated by different IoT medical sensors and send it to the edge (fog) for preprocessing. The useless data/attributes should be cleaned/removed.
2. Data filtering/classification focusing on classifying the generated data based on their privacy level:

    a. Normal data (non-sensitive): This type of data is related to the device information which includes id, serial number, model, browser name, version, etc. Moreover, it may include network details and cookies.
    b   Health-critical data (private): This type of data privacy is related to health-related data. It is more critical and any attack on these data may expose its users to risky issues. The preprocessed data can be classified as follows:

        • Abnormal-health data: this data indicated risky biomarkers measurements that should ensure accuracy and that is because the disease diagnosis will depend on these measurements.
        • Normal health data indicates the usual health status and normal biomarkers.

    c   Private (sensitive), this data privacy category includes user's personal information e.g., name, age, gender, account password, etc.

3. Send notification to doctor and patients in case of abnormal measurements or readings.

    In case of abnormal readings, the message as a serious notification will be sent to the doctor and patient. In this case, the doctors can access all patient's related data for diagnosis purposes. Therefore, access to the data in the fog will be restricted only to authorized doctors. Thus, accessing patients' data in fog will be restricted only to authorized doctors.

4. Send normal data to the cloud.

    The data temporarily stored in fog servers two and three will directly be sent to the cloud for permanent storage and analysis. The critical data, which are stored in fog server one will be processed in case of abnormal measurement, and after diagnosing the case, the data will be sent to the cloud for permanent storage.

6 Discussion

As stated previously, processing and computing the data in the edge (fog) nearby IoT-based medical devices help to solve issues related to the response time, delay, and network bandwidth [27]. The data generated by IoT medical sensors can be preprocessed in the edge servers (fog) for reliability and efficiency enhancement in IoT healthcare [28]. The focus of the work is the data privacy issue. Thus, the proposed approach targets the fog for data privacy preservation for data generated by

IoT medical devices. The proposed work only illustrated the scenario that shows the process of transferring the data from IoT medical sensors to the gateway and edge servers, ending with cloud storage.

According to the data types generated by IoT medical sensors, the data can be categorized into three categories based on their privacy level and criticality. The first category is the normal data which is non-private such as device information. The second category is private data which is more critical and requires more protection. The third category is related to the private sensitive data which is not critical such as personal information but also requires protection.

The proposed approach focuses on preserving privacy after preprocessing the data. More precisely, the preprocessing involves classifying or filtering the data into several fog servers which are dedicated to the data categories which have been identified before. The data preprocessing will help to filter the data across different servers considering the category type of data. For example, critical data such as blood pressure measurement will be sent to the fog server dedicated only for computing such data. This operation of disseminating data across different servers will help to apply advanced data protection to only fog servers that hold the more critical data.

Moreover, the classified data which are more critical can be processed faster and this will be useful especially in case of discovering any abnormalities. This will help to reduce the latency and thus improves the diagnosis process. The proposed approach also supports notification features in which if abnormalities are detected, the notification will fire to the doctors and the patients. While in another case, if everything is normal, the data will be transmitted to the cloud servers for further analysis. The doctors also have the access to the patient's data in the edge. Advanced artificial intelligence methods can be applied for implementing this proposed approach. For example, the Support Vector Machine (SVM) can be used as a classification method for classifying the data as normal and abnormal. SVM can be implemented in fog due to its low computation cost compared to other machine learning algorithms [27].

7 Conclusion and Future Work

Edge (fog computing) is a new development that mimics the cloud in many functions, and it comes with many solutions to cloud issues in IoT applications such as high network bandwidth, latency, and response time. In the healthcare scenario, IoT applications of real-time monitoring require fast data computation with low latency and high response time. The success of such application in clinical practice will help in improving the quality of diagnosis of diseases. However, due to the massive data generated by IoT medical sensors with its variability in sensitivity level, data privacy remains the major obstruct to employ these devices in hospitals. Therefore, there is a need for a data privacy approach that can assist in protecting the data especially the most critical data by ensuring a high response time and low latency. Thus, the proposed scenario of data privacy in the fog provides a clear picture of the streamed data in the fog servers and its classification and protection.

The streamed data which are generated by IoT medical sensors can be disseminated in fog virtual servers into three main categorized (non-private, critical private, non-critical private). The machine learning classification algorithms can be used for data classification and advanced data protection algorithms can be applied only for the most critical data. The non-private data will be sent to the cloud for further analysis. So, the fog servers will process the generated data faster than the cloud because it is located nearby the IoT medical devices and this will help in detecting the abnormalities in patients in a faster way. It will be also easier to build robust reports from the cleaned data which is the critical one. This work will be extended for real implementation and evaluation.

Acknowledgements This work is funded by Omantel under the project code [EG/SQU-OT /1802]. This work is as a part of project title "Internet of Things (IoT) security and privacy aspects related to architecture, connectivity, and collected data".

References

1. Alshohoumi, F., Sarrab, M., Al-Abri, D., Al Hamadani, A.: Systematic review of existing IoT architectures security and privacy issues and concerns. Int. J. Adv. Comput. Sci. Appl. (IJACSA) **10**(7), 232–251 (2019)
2. Alhazmi, O.H., Aloufi, K.S.: Fog-based internet of things: a security scheme. In: 2019 2nd International Conference on Computer Applications & Information Security (ICCAIS), pp. 1–6 (2019). https://doi.org/10.1109/CAIS.2019.8769506
3. Sood, S.K., Mahajan, I.: IoT-fog-based healthcare framework to identify and control hypertension attack. IEEE Internet Things J. **6**(2), 1920–1927 (2019). https://doi.org/10.1109/JIOT. 2018.2871630
4. Macdermott, A., Kendrick, P., Idowu, I., Ashall, M., Shi, Q.: Securing things in the healthcare internet of things. In: 2019 Global IoT Summit (GIoTS) (2019). https://doi.org/10.1109/GIOTS. 2019.8766383
5. Alharam, A.K., El-Madany, W.: Complexity of cybersecurity architecture for IoT healthcare industry: a comparative study. In: 2017 5th International Conference on Future Internet of Things and Cloud Workshops (FiCloudW), pp. 246–250 (2017). https://doi.org/10.1109/FiC loudW.2017.100
6. Baker, S.B., Xiang, W., Atkinson, I.: Internet of things for smart healthcare: technologies, challenges, and opportunities. IEEE Access **5**, 26521–26544 (2017). https://doi.org/10.1109/ ACCESS.2017.2775180
7. Pulkkis, G., Westerlund, M., Karlsson, J., Tana, J.: Secure and reliable internet of things systems for healthcare, pp. 169–176 (2017). https://doi.org/10.1109/FiCloud.2017.50
8. Ozcan, K., Velipasalar, S., Varshney, P.K.: Autonomous fall detection with wearable cameras by using relative entropy distance measure. IEEE Trans. Hum.-Mach. Syst. (2017). https://iee explore.ieee.org/iel7/6221037/6340045/07740939.pdf. Accessed 13 Jan 2019
9. Islam, S.M.R., Kwak, D., Kabir, M.D.H., Hossain, M., Kwak, K.-S.: The internet of things for health care: a comprehensive survey. IEEE Access **3**, 678–708 (2015)
10. Yeole, A.S., Kalbande, D.R.: Use of internet of things (IoT) in healthcare, pp. 71–76 (2016). https://doi.org/10.1145/2909067.2909079
11. Torre, I., Koceva, F., Sanchez, O.R., Adorni, G.: A framework for personal data protection in the IoT. In: 2016 11th International Conference for Internet Technology and Secured Transactions (ICITST), pp. 384–391 (2017). https://doi.org/10.1109/ICITST.2016.7856735

12. Nahapetian, A.: Side-channel attacks on mobile and wearable systems. In: 2016 13th IEEE Annual Consumer Communications & Networking Conference (CCNC), pp. 243–247 (2016)
13. Bahirat, P., He, Y., Menon, A., Knijnenburg, B.: A data-driven approach to developing IoT privacy-setting interfaces. In: 23rd International Conference on Intelligent User Interfaces, pp. 165–176 (2018)
14. Sarrab, M., Alshohoumi, F.: Privacy concerns in IoT a deeper insight into privacy concerns in IoT based healthcare. Int. J. Comput. Digit. Syst. **9**(3), 399–418 (2020)
15. Chen, D., Kalra, S., Irwin, D., Shenoy, P., Albrecht, J.: Preventing occupancy detection from smart meters. IEEE Trans. Smart Grid **6**(5), 2426–2434 (2015)
16. Barker, S., Kalra, S., Irwin, D., Shenoy, P.: Powerplay: creating virtual power meters through online load tracking. In: Proceedings of the 1st ACM Conference on Embedded Systems for Energy-Efficient Buildings, pp. 60–69 (2014)
17. Chen, D., Bovornkeeratiroj, P., Irwin, D., Shenoy, P.: Private memoirs of IoT devices: safeguarding user privacy in the IoT Era. In: 2018 IEEE 38th International Conference on Distributed Computing Systems (ICDCS), pp. 1327–1336 (2018). https://doi.org/10.1109/ICDCS.2018.00133.
18. Djenna, A., Saidouni, D.E.: Cyber attacks classification in IoT-based-healthcare infrastructure. In: 2018 2nd Cyber Security in Networking Conference (CSNet), pp. 7471–7474 (2019). https://doi.org/10.1109/CSNET.2018.8602974
19. Andriopoulou, F., Dagiuklas, T., Orphanoudakis, T.: Integrating IoT and fog computing for healthcare service delivery. In: Components and Services for IoT Platforms, pp. 213–232. Springer (2017)
20. Gill, S.S., Arya, R.C., Wander, G.S., Buyya, R.: Fog-based smart healthcare as a big data and cloud service for heart patients using IoT. vol. 1, pp. 1376–1383 (2019). https://doi.org/10.1007/978-3-030-03146-6_161
21. Farahani, B., Firouzi, F., Chang, V., Badaroglu, M., Constant, N., Mankodiya, K.: Towards fog-driven IoT eHealth: promises and challenges of IoT in medicine and healthcare. Future Gener. Comput. Syst. **78**, 659–676 (2018)
22. Muniz, J., Lakhani, A.: Investigating The Cyber Breach: the Digital Forensics Guide for the Network Engineer. Cisco Press (2018)
23. Alrawais, A., Alhothaily, A., Hu, C., Cheng, X.: Fog computing for the internet of things: security and privacy issues. IEEE Internet Comput. **21**(2), 34–42 (2017). https://doi.org/10.1109/MIC.2017.37
24. Elgendy, I.A., Zhang, W., Tian, Y.C., Li, K.: Resource allocation and computation offloading with data security for mobile edge computing. Future Gener. Comput. Syst. (2019). https://doi.org/10.1016/j.future.2019.05.037
25. Abou-Nassar, E.M., Iliyasu, A.M., El-Kafrawy, P.M., Song, O.Y., Bashir, A.K., El-Latif, A.A.A.: DITrust chain: towards blockchain-based trust models for sustainable healthcare IoT systems. IEEE Access (2020). https://doi.org/10.1109/ACCESS.2020.2999468
26. Elgendy, I.A., Muthanna, A., Hammoudeh, M., Shaiba, H., Unal, D., Khayyat, M.: Advanced deep learning for resource allocation and security aware data offloading in industrial mobile edge computing. Big Data (2021). https://doi.org/10.1089/big.2020.0284
27. Sarrab, M., Alshohoumi, F.: Assisted-fog-based framework for IoT-based healthcare data preservation. Int. J. Cloud Appl. Comput. (IJCAC) **11**(2), 1–16 (2021)
28. Azimi, I., et al.: HiCH: hierarchical fog-assisted computing architecture for healthcare IoT. ACM Trans. Embed. Comput. Syst. **16**(5s), 1–20 (2017)

Trusted Execution Environment-Enabled Platform for 5G Security and Privacy Enhancement

José María Jorquera Valero, Pedro Miguel Sánchez Sánchez, Alexios Lekidis, Pedro Martins, Pedro Diogo, Manuel Gil Pérez, Alberto Huertas Celdrán, and Gregorio Martínez Pérez

Abstract With the deployment of 5G networks and the beginning of the design of beyond 5G communications, new critical requirements are emerging in terms of performance, security, and trust for leveraged technologies, such as Software Defined Networking (SDN) and Network Function Virtualization (NFV). One of the requirements at the security and trust level is that when delegating critical tasks and data to the infrastructure deployed in an external domain, the client needs guarantees that the execution has been carried out securely, without data breaches or compromises during computing tasks. To meet this need, this chapter proposes a framework that uses Trusted Execution Environments (TEEs), processing environments isolated from the rest of the system to guarantee the security of the data and tasks processed in them, in order to improve the security of 5G environments. This framework enables the deployment of TEE as a cloud service, also denoted as TEE-as-a-Service or TEEaaS,

J. M. J. Valero (✉) · P. M. S. Sánchez · M. G. Pérez · A. H. Celdrán · G. M. Pérez
Department of Information and Communications Engineering, University of Murcia, 30100
Murcia, Spain
e-mail: josemaria.jorquera@um.es

P. M. S. Sánchez
e-mail: pedromiguel.sanchez@um.es

M. G. Pérez
e-mail: mgilperez@um.es

A. H. Celdrán
e-mail: alberto.huertas@um.es

G. M. Pérez
e-mail: gregorio@um.es

A. Lekidis
Intracom Telecom, 190 02 Athens, Greece
e-mail: alekidis@intracom-telecom.com

P. Martins · P. Diogo
Ubiwere, 3800-075 Aveiro, Portugal
e-mail: pmartins@ubiwhere.com

P. Diogo
e-mail: pdiogo@ubiwhere.com

© The Author(s), under exclusive license to Springer Nature Switzerland AG 2022
A. A. Abd El-Latif et al. (eds.), *Security and Privacy Preserving for IoT and 5G Networks*,
Studies in Big Data 95, https://doi.org/10.1007/978-3-030-85428-7_9

allowing customers to take advantage of its benefits without having to deal with the configuration of the environment and hardware. Furthermore, this chapter also discusses current trends as well as future challenges related to the deployment of TEEs in 5G environments, providing key aspects for future solutions in the area.

Keywords 5G cyber-security · Trusted execution environments · Multi-domain environments · Trustworthy network slicing · Cloud infrastructure

1 Introduction

The modern network paradigm is enabling 5G developments where millions of network elements are connected in real time with scalability and performance far superior to previous versions of the network infrastructure (i.e., 3G and 4G) [20]. In this line, virtualization technologies are emerging as the main drivers for service and resource management due to their flexibility in configuration and deployment. In addition, network resources forming a service can be distributed across domains in different organizations. This situation enables a rapidly changing environment in which network services are created and distributed very dynamically [28].

In this context, there is an ongoing trend for some organizations to offer or share computing services by renting resources that are surplus or close to the end user, offering a better user experience. This allows optimizing the use of resources while offering a better Quality of Service (QoS) to end users, getting faster responses to requests and not exposing information beyond the end user's environment, for example. However, offloading tasks on an external organization infrastructure, especially when they are critical tasks or sensitive data management [34], implies new security risks since direct control of the infrastructure where the software is running is lost. This situation implies additional security measures to ensure reliable execution while maintaining good efficiency and revenue.

To solve the problem, several actions can be taken, such as the use of encrypted communications [15], correct data life cycle handling [59], etc. In addition, some measures are also necessary to guarantee the isolation and security of the processes during their execution. In this sense, the deployment of Trusted Execution Environments (TEEs) becomes one of the most relevant options to guarantee the security of the critical processes executed [61].

A TEE is an isolated processing environment where the code and the data are protected during execution, decoupling its memory area from the rest of the processor and providing confidentiality and integrity properties [49]. To provide such capabilities, the environment should run on a separated kernel whose components (CPU registers, memory, sensitive I/O, etc.) are trustworthy against software and physical attacks. Nevertheless, the usage of TEE in real world deployments comes with given associated problems such as specific code and software adaptations and compiling, specific TEE configuration, or performance degradation [43]. Besides, the integration of these solutions into modern scenarios, such as 5G deployments, brings additional challenges due to the dynamicity and quick deployment time required.

Then, despite the great advances in TEE and 5G virtualization technologies, there are still relevant challenges to tackle, mainly related to their integration. Among them, we highlight the following ones:

- **Application of TEEs on top of existing software solutions**. The fact of adapting pieces of software already developed and compiled to be executed in TEE solutions can be a complex task depending on the used TEE framework.
- **5G core network component execution in TEE frameworks**. 5G network elements have critical performance requirements in terms of throughput and delay, but also security. Then, it is critical to find a proper balance between security guarantees and service performance.
- **TEE offered as a service for offloaded task execution**. Traditionally, TEE solutions are deployed internally in each domain, without offering these capabilities to potential external customers. However, TEEs offer advanced security and isolation capabilities that, provided as a cloud service, can generate a new service market, just like infrastructure (IaaS), platforms (PaaS), and software (SaaS).

With the goal of improving the security in offloaded task execution on demand, this book chapter proposes a TEEaaS framework for 5G multi-domain networks in which multi-tenancy infrastructures are considered as a key enabler. The proposed framework seeks to guarantee trusted execution of code when offloading critical data and processes into a foreign network domain due to possible performance and system requirements. This framework seeks to decouple possible TEE configuration and deployment tasks from its final usage in actual code, facilitating its use in real environments and generating a new service of security and trust. Thus, the TEEaaS leverages SCONE [9] as a mechanism for abstracting specific implementation details. The present work is developed under the 5GZORRO H2020 project [27], which focuses on the development of solutions for zero-touch service, network, and security management in multi-stakeholder environments operating in a 5G context. Then, the ultimate goal of the proposed framework is to provide TEEaaS solutions in a marketplace shared among several tenants that maintain a commercial relationship based on networked services and resources, where consumers have the possibility to instantiate services or applications on a set of resources that may be, by nature, unreliable.

The rest of the chapter is structured as follows. Section 2 provides the required background on 5G network architecture and TEE technologies, which are useful for the comprehension of the proposed approach, and in particular, Sect. 2.3 that introduces the current state-of-the-art about TEEs and 5G enforcements. Section 3 details the proposed solution for a TEEaaS platform, focusing on its design and implementation. Then, Sect. 4 provides an insight into current trends and future challenges on TEE application in modern scenarios. Finally, Sect. 5 summarizes the presented solution as well as provides some perspectives for future work.

2   Background and Related Work

In order to understand in depth our TEE-enable solution, it is critical to have a comprehensive view of the 5G paradigm and the main technologies leveraged in it. Besides, it is equally important to provide a background in TEE technologies, with a high-level description and definition of the main types of TEEs. Thus, this section provides the background insights required regarding 5G and TEE technologies.

2.1   5G Network Architecture and Enablers

5G is expected to be a multi-tenant and multi-service network where technologies play a paramount role as a pillar of digital evolution. Simultaneously, the growth of interconnected devices on 5G networks entails more efficient use of mobile network infrastructures, as well as ensuring enhanced bandwidth, ultra-low latency, network coverage, and high data capacity compared to previous networks. Thus, 5G is designed thinking in a dynamic network environment, where networks are created and modified according to heterogeneous requirements. In this scope, network slices are considered as a necessary mechanism for satisfying the QoS requirements as well as enabling the coexistence of multiple verticals [29].

To solve the previous requirements, advanced networking and virtualization methods, such as Network Function Virtualization (NFV) and Software Defined Networks (SDNs), are regularly contemplated in literature. Thus, flexible virtualization approaches, also known as Virtualized Network Functions (VNFs), are employed as the adoption of techniques that provide not only a high QoS but also a low cost in network environments, thus minimizing operational and capital expenditures (OPEX and CAPEX). Despite the benefits that multi-tenant and multi-service network slices bring, they also introduce a potential attack surface for 5G networks [21]. In this sense, adversaries might eavesdrop on network communications, gain unauthorized access, and finally, carry out spiteful actions such as disrupting swapping data by users or tenants.

In the case of NFV, it possesses a complex architecture [46] that makes it prone to cyber-attacks. In particular, the Virtualized Infrastructure Manager (VIM), which is one of the three functional blocks of ETSI NFV Management And Network Orchestration (NFV-MANO) [44], is a pivotal component as well as one of the most targeted. Among the main tasks of VIM is to coordinate the NFV Infrastructure (NFVI) resources utilized by the VNFs, based on what is needed at any given time, to deliver network services and resources. Therefore, adversaries may discover vulnerabilities that would allow them to steal or tamper with sensitive data [40] and critical processes running within the VIM. To dwindle these types of attacks, the VIM might consider the use of trusted platforms to perform certain critical actions [53]. For the sake of illustration, a possible countermeasure could be the VIM trusted execution by deploying these as containerized systems isolated from the host system. Thus, the

VIM life cycle would be executed inside trusted platforms, and therefore, the boot processes, connections and data management, kernel memory, and code integrity may evade buffer overflow and code injection attacks.

Regarding SDN, it encompasses a wide variety of architectures and functionalities such as controllers, northbound and southbound APIs, SDN applications, etc. SDNs promote a dynamic, programmatically, and efficient network administration using software. Nevertheless, the variety of architectures and functionalities together with high customization of the networks lead to the emergence of new security risks in 5G networks. Multiple security threats may be linked to SDNs, among the most well-known we can introduce data forging, memory scripting [7], or DDoS [23, 30], to name but a few. On the one hand, the data forging entails compromising an SDN element such as a switch, router, or controller to falsify network data and launch other attacks as DoS. In this way, data forging has been identified as a threat related to components in the data plane and the controller plane. On the other hand, the memory scripting threat is carried out by an adversary scanning the physical memory of a software component to acquire sensitive data for which he/she/it does not have permission. These threats may cut down the SDN security risks through a trusted executed platform that guarantees not only code integrity to prevent memory or file modification (against memory scripting attacks), but also code confidentiality to ensure code loading (against data forging attack) [42].

Another pivotal enabler that has been boosted by the 5G ecosystem is Distributed Ledger Technologies (DLTs). Having in mind heterogeneous and multi-stakeholder 5G ecosystems, such as a trustworthy end-to-end establishment across multiple domains [27], DLTs facilitate the interaction among multiple stakeholders involved in dynamic 5G environments, the cross-domain sharing of crucial data (Smart Contracts), accelerate settlements that previously involved complex processes, and the decentralized management of the security-related information [35]. In this scope, DLTs are envisioned to play a crucial role in 5G scenarios. Nevertheless, such technology does not depend on pre-established trust between the implicated entities, therefore it also introduces challenges that need to be addressed such as improving the security and the privacy of data being pushed into the DLT or ensuring the nodes against data leakage and injection attacks. In particular, the use of a trusted environment may ensure that sensitive operations, such as the Service Level Agreement (SLA) computation monitoring [22] or authenticity proofs for smart contracts can run inside a tamper-proof environment [58], where neither the off-chain data nor its computation can tamper, and consequently avoiding possible data leakage and injection attacks.

As can be appreciated from the previous examples, there are multiple types of threats affecting swapping, confidentiality and integrity data, or disrupting the normal life cycle of resources. In that sense, 5G network components and enablers need to diminish the attack surface as well as solving or mitigating the feasible threats. Additionally, in the 5G multi-tenant and multi-service environments, these threats may display a higher risk since they may trigger lateral movements targeting other tenant resources. For that reason, we contemplate the use of TEEs as reliable service platforms that enable ensuring [54] the confidentiality and integrity of data managed in distributed, dynamic, and flexible environments, as well as ensuring the non-interruption of the processes carried out in our structures.

2.2 TEE Background

TEE is a technology used to provide a tamper-resistant processing environment that runs on a separation kernel. Such a kernel enables systems with different levels of security to coexist on the same platform. The TEE can resist both software attacks and physical attacks performed on the main memory of the system. Furthermore, attacks performed by exploiting backdoor security flaws or system vulnerabilities are also avoided using TEE [49].

With respect to its functionality, the TEE divides the system into trusted and untrusted partitions that are isolated. These partitions use a secured interface for inter-partition communication to ensure that no lateral movement can occur if the untrusted partition is compromised. Currently, the secure interfaces are implemented using three main mechanisms: GlobalPlatform TEE Client API [3]; secure Remote Procedure Call (RPC) of Trusted Language Runtime [51]; and real-time RPC of dual-OS system [50].

Furthermore, three categories of TEE implementations are usually found: at the hardware level, at the software level, and integrated hardware-software solution. Hardware-based TEEs are using enforced isolation that is built into the CPU. A characteristic example is the Arm TrustZone technology [60]. Extensions to this technology allow securing the memory area. On the one hand, a hardware-based TEE allows trusted applications to have complete access to the main processor, resources (e.g., peripherals as sensors, actuators), and memory, while hardware isolation protects them from untrusted applications running on the main operating system.

On another hand, software-based TEEs are providing a normal and secure partition at the device firmware level, where both the kernel and the operating system are encrypted with strong cryptographic mechanisms to protect the data and applications of each device. Access to the keys for encrypting/decrypting the data is provided only to trusted entities through proper authentication or authorization schemes. A characteristic example of software-based TEE is the Open Portable Trusted Execution Environment (OP-TEE) [6]. The main difference with hardware-based TEEs is that they do not provide protection schemes for hardware resources, hence applications may only rely on the trusted kernel and not the processor and resources.

Concerning the integrated hardware-software TEE solution, it is formed as a combination of the hardware- and software-based TEEs to provide protection with multiple security layers. These layers ensure strong encryption mechanisms for applications, system configurations as well as stored data at both hardware and software (i.e., operating system) level.

Several open-source and proprietary TEE platforms are described in the literature as well as are available as market or community solutions. These implementations are separated into the three aforementioned categories as follows:

1. **Hardware-based TEEs**

   - *Arm's TrustZone* (proprietary) [16] offers an efficient system-wide approach to security with hardware-enforced isolation built into the CPU. Genode TEE [1] is based on this technology and extends it through a hypervisor environment that allows switching between secure and normal world.
   - *Intel SGX* (proprietary) [32] provides hardware-based memory encryption that isolates data and application-specific code in memory. Intel SGX allows user-level code to allocate private regions of memory, called *enclaves*, to protect against processes running at higher privilege levels. Besides, it provides a granular level of control and protection.
   - *Intel Trusted Execution Technology (TXT)* (proprietary) [4] provides a root-of-trust and verifies the integrity of a platform by relying on a TEE. During boot, it performs measurements on the platform components (boot loader, firmware, hypervisor, operating system) and verifies them against pre-calculated whitelist values. Hence, it provides mechanisms to protect vital data and processes from being compromised by possible malicious software running on the platform.

2. **Software-based TEEs**

   - *OP-TEE* (open-source) [6] is a TEE solution designed as a companion to a non-secure Linux kernel running on Arm. However, it has been structured to be compatible with any insulation technology suitable for the TEE concept and objectives: isolation, small footprint, and portability.
   - *Trusted Little Kernel* (open-source) [13] is a TEE solution by NVIDIA which defines a software-partitioned environment that provides trusted operations and supports multi-threading, a monitor for switching between the secure and normal environment and finally, a secure storage.

3. **Integrated TEEs**

   - *Trustonic* (proprietary) [12] allows devices to be embedded with a Trusted Identity, combining a secure OS (kernel and memory) along with a hardware-secured environment (e.g., Arm TrustZone chip). Trustonic is focused on mobile, automotive, banking, and IoT sectors and provides key features such as secure boot, secure data storage, secure execution, protecting connected hardware, and secure channels for application delivery.
   - *Solacia SecuriTEE* (proprietary) [11] is a hardware environment that provides security service for processors, peripherals, and storage devices, which is running on top of an Arm TrustZone chip. Additionally, the hardware environment is coupled with a secure kernel, which contains a trusted core environment, trusted function, and an integrated API for communication with the untrusted partition. The API adheres to the GlobalPlatform TEE mechanism.

An overview of the TEE architectural deployment is illustrated in Fig. 1. This figure builds on top of existing TEE functional views [49] and extends it with the modules that are included in the existing TEE implementations. The figure depicts an untrusted as well as a trusted area, which communicates using the secure API.

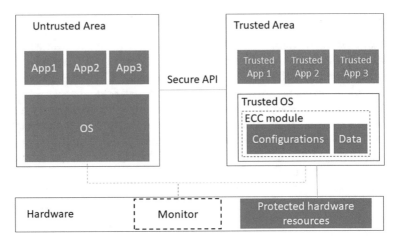

Fig. 1 Trusted execution environment overview

The Trusted Area uses kernel encryption and Elliptic Curve Cryptography (ECC) primitives to protect the data as well as the application and system configurations inside the operating system. Furthermore, trusted applications are meant to handle confidential information such as credit card PINs, private keys, and Digital Rights Management (DRM) protected media, among others, as well as providing services to the normal world OS to make use of confidential information without compromising it. Finally, dedicated hardware is used to protect the associate hardware resources in addition to enabling optional capabilities for switching between a secure and normal (i.e., untrusted) environment.

The offered security level of hardware-based and integrated TEE solution is significantly higher than a software-based TEE, mainly due to the higher safety that is offered by hardware encryption mechanisms. On the other hand, hardware-based encryption usually requires dedicated hardware, which results in a significant cost increase for such solutions. The advantages and disadvantages of each of the presented TEE categories towards the design of a TEEaaS platform are further described in Sect. 3.

2.3 Related Work on TEEs Application in 5G

Now, we review the main works combining TEE technologies with 5G and its enablers. Although 5G is a novel technology, there are already some works leveraging TEEs in its networking and solution context.

Ortiz et al. [47] considered TEEs as a game-changing technology in the security of virtualized environments in 5G networks, mainly in integrity and confidentiality perspectives. Here, TEEs are envisioned as a solution for virtual machine or container isolation mechanisms, preventing introspection attacks into the host machine. In [33], Elbashir analyzed how TEEs can be integrated with SDN solutions as a security enabler for 5G mobile networks. Concretely, this work proposed the usage of Intel SGX for TLS connection management (TLSonSGX) between switches deployed using Open vSwitch. In turn, the authors of [42] proposed a TEE abstraction layer for AMD's SEV and Intel's SGX which enables the secure deployment of SDN and NFV solutions. Similarly [19], considered TEE for VNF (and its hypervisor) isolation and enhancing the security.

From a networking standpoint, Kim et al. [41] proposed the usage of TEEs for enhancing the security and privacy of a confidential networking solution. This approach was designed, deployed, and validated using the TOR's (The Onion Router) networking ecosystem. Besides, the idea of using TEEs for data protection in collaborative networks is explored in [25] in a vehicular network system context. From another perspective, the security architecture for 5G presented in [17] mentioned the TEE possible application in user equipment as a hardware secure layer for certificate and credential management.

As it can be seen, there are many works leveraging TEE technologies in the 5G context. However, none of them envisions the usage of TEEs as a service, offering to the customers additional trust and security properties for their services.

3   Design of the TEEaaS Platform

A distributed and virtualized 5G network in a multi-stakeholder and multidomain third-party infrastructure creates a scenario where no inherent trust mechanisms can exist. Therefore, built-in security is paramount at all operational levels, from the hardware to the virtualized layers, but also covering both data and software. TEE-based software execution enables critical workloads to go across different tenants and stakeholders with no losses in security, by providing an isolated processing environment.

In this section, we start detailing how the usage of hardware-based TEEs provides a secure environment for deploying network components, such as VIMs and VNFs, in a third-party infrastructure. Later, we focus on the development of such functionalities by integrating commercial TEEs in the execution of some 5G network virtualized software components, enhancing the security and trust of the software executed under these capabilities. Besides, we also present a TEEaaS solution for 5G multi-domain networks inspired by the 5GZORRO H2020 project [27]. Finally, we end this section by presenting how such secure systems with TEEs can be designed to provide a trustworthy execution environment for 5G applications.

3.1   Secure Environments for Critical Workloads

Protecting a tenant's service or application running on a third-party computing node against a malicious entity, with root and/or physical access, requires a root-of-trust built-in from the hardware. Zero trust hardware platforms, where both data and code are protected even from the infrastructure owner, enforce the necessary security to protect the application from two scenarios:

1. A malicious stakeholder wants to access/manipulate the data or code from a tenant running on its physical infrastructure.
2. A malicious third-party exploits vulnerability in the infrastructure to inadvertently access the computation node which is running the virtualized network modules and tampers with the execution.

TEE-based software execution ensures that only the code running inside the TEE can access its own data. Therefore, critical operations can be executed independently if the system, either hardware, software, or both, has been compromised. The separation between trusted and untrusted areas allows a malicious element to exploit security flaws in software without compromising the critical data and critical operations.

Since a TEE includes a zero trust hardware platform, a key component to establish a root-of-trust and end-to-end secure communications. the presence of these capabilities in the infrastructure offered by some stakeholders also improve the trust level perceived by service consumers, providing extra security at the hardware level to the resources and services offered by service providers.

TEEs are not a new technology by themselves. They are already mainstreamed in contactless payments and biometrics scans on mobile devices. In these scenarios, they enforce a secure enclave: a hardware-enforced separation between the untrusted and the trusted area. This separation allows a third-party application, service, or task to perform operations using critical data without exposing its access to this data or device, nor introducing attack vectors due to security flaws in its design.

The novelty element of the application of TEEs is to expand this concept of secure enclaves from mobile devices to a 5G telecommunications infrastructure, enabling two new use cases:

1. Deployment of network components such as VNFs, VIMs, and orchestration services in a third-party infrastructure [37].
2. Execution of secure oracles, DLTs nodes, authenticity proofs for smart contracts, and other critical operations, where neither the off-chain data nor its computation can be tampered [45, 55].

In the same way that mobile applications thrived when application marketplaces were introduced by mobile device manufacturers, distributed and virtualized telecommunications infrastructure is expected to thrive when similar marketplaces are introduced for VIMs, VNFs, and NFVs [24].

3.2 Suitable TEEs Based on Cloud Infrastructure

In this section, we describe the applicable solutions for the cloud infrastructure-related TEEs, which are built on top of the solutions explained in Sect. 2 to separate secure from non-secure areas and create an isolated processing environment. These solutions are comprised of hardware, software, or both and can either run x86, Arm, or RISC-V Instruction Set Architecture (ISA).

While IoT devices can be a part of a telecommunication network [38], an Arm-based TEE suffers from a shortcoming: when compared with x86 or RISC-V based TEEs, they can only have a single secure environment per processor. In a distributed and shared infrastructure, which supports concurrent execution and 5G physical medium access, it is desired that each computation node supports the instantiation of multiple, concurrent secure enclaves, so that multiple tenants can concurrently have a secure enclave allocated to themselves. While RISC-V based microprocessors and microcontrollers have been gaining traction in the last years [5], there are not yet mainstreamed, nor their TEEs modules are mature enough so that we can consider them as a viable commercial solution when compared to x86 solutions.

Therefore, in the context of a virtualized 5G network, x86-based TEEs will still dominate in the forthcoming years. It is worth noting that, due to the virtualization component of networks, the computation node that supports a TEE does not require to be near the edge, therefore, current cloud infrastructure can be used. This realization makes x86 the preferred TEE solution, due to its dominance in the Cloud-as-a-Service market. Since most of the market share for desktop and cloud processors is dominated by Intel, its TEEs are the most prevalent and available cloud-based solutions.

3.3 TEEaaS

Despite market trends, a solution for multi-domain, multi-stakeholder, and distributed 5G networks should be vendor-agnostic and independent of the TEE solution to which the infrastructure has access, as long as the infrastructure can guarantee certain capabilities. Due to the virtualized nature of the infrastructure, abstracting the low-level details of the commercial TEEs available and exposing only high-level tasks to any service or application is not only desired but necessary. By implementing an API for "TEEaaS" to allow the execution of modules in a secure enclave on the distributed 5G network:

- The know-how required to operate with hardware-based TEEs for secure enclaves is not required by the network engineers deploying the containerized network modules.
- New TEE solutions can be added seamlessly to the network infrastructure by different providers, without requiring the upper-level software modules to be ported to different hardware.

Besides, the TEEaaS platform should also offer capabilities to enable the execution of other simpler components and code, such as tasks manipulating critical data, improving the flexibility of the objects that can be deployed leveraging the TEE properties. In the scope of the 5GZORRO H2020 project and other 5G networks, containerized modules can be integrated with the orchestration services, allowing the deployment of critical services on TEE-enabled nodes present in the marketplace. In this regard, SCONE - Secure Linux Containers for Confidential Computing [9] is a framework that has been initially developed in the context of different H2020 projects (mostly Sereca [8] and Secure Cloud [10], while others have been exploiting it and enhancing it [9]). SCONE abstracts specific implementation details of the Intel's SGX secure enclave, but also provides encryption at rest, in transit, and during runtime without requiring source code changes, supporting most modern program languages. It also has built-in attestation and key provisioning modules, allowing the application developers to focus on the orchestration and configuration of the security management solution and not on the security-solution implementation.

With a focus on the SCONE framework to enable TEE capabilities, the 5GZORRO H2020 project has adopted a hardware-based TEE approach, specifically Intel's SGX, in order to provide a TEEaaS solution for 5G multi-domain networks. When it comes to the instantiation of applications which are ready to be used in a TEE environment (i.e., Intel SGX), this functionality has not been altered, but it is now offered following a Cloud Native way. In other words, such services need only to be instantiated using Kubernetes or Docker with SCONE abstracting the interface within Intel SGX, and therefore, not requiring the definition of specific APIs. Furthermore, from a design point of view, there is no need to change the interface with other components, nor the high-level functional capabilities.

Hence, TEEaaS is expected to enhance 5GZORRO's ability not just to run 5GZORRO core components in a TEE environment (as an SLA monitoring service), but most importantly, the ability to offer consumers the capability to instantiate services in a pool of resources which may be untrusted by nature. For instance, a single and centralized instance of such a component will be enough to feed all subsequent TEE-powered NFVI for attestation and configuration purposes.

3.4   From TEEaaS to Full Application Security

Integrating the TEEaaS with the current ETSI NFV MANO tools for 5G networks requires not only that the orchestrating services can deploy a custom application in a secure enclave, but also that the deployment of the application is performed ensuring end-to-end encryption and secure provisioning of the application, its data, and keys. 5G networks must protect data and software while they are running on the secure enclave, but also while data and code are in transit and at rest.

While multiple secure enclaves and TEE implementations are available, the TEEaaS is used to create a common layer for such secure enclaves. In this scope, a secure enclave on its own is only part of the solution to achieve a complete application-oriented security solution. Moreover, the TEEaaS allows the data and applications to be secure in all system states: during runtime, at rest, and in transit.

Such capabilities require that the application data and files must be encrypted such that they can only be accessed by the application itself. Hence, a TEE-based application should have access to the decryption keys, which implies the following steps:

1. An external module to the application must be responsible for managing the secrets and keys of the application. This module should also be secured and capable of provisioning the secrets to a genuine application securely.
2. The application and data must be verified and prove to be genuine, i.e., that its code and data have not tampered with.
3. The tenant must be able to verify and attest that the host provided a tamper-free environment (secure enclave inside a TEE) for the tenant to run the application.
4. Secure, end-to-end encrypted communication must be established between the external module described in step 1 and the genuine application, running in a previously attested environment.

The above points must be fulfilled by the orchestration services, in order to ensure that the applications and corresponding data for a 5G network module are secure in all phases of its life cycle.

4   Current Trends and Future Challenges

Given the demonstration of the main types of TEE solutions along with their prevalent properties and limitations, in this section, we focus on the trends and limitations of their offered mechanisms as well as of their adoption on 5G multi-domain environments. These trends reflect the path that the future TEE implementations will have to follow in order to, on the one hand, fill the current gaps associated with performance and integration issues, and on the other hand, meet the security and privacy-preserving challenges of such environments.

4.1   Current Trends

A current trend towards the TEE adoption in 5G multi-domain environments is its integration with the VIM or NFVI components of the ETSI NFV MANO architecture [44]. One possible approach in this direction is provided by [53], which describes that a VIM with trusted execution capabilities is achieved by allowing isolation, both at the hardware and the software level for controller nodes. Isolation allows protecting the sensitive data of applications and services that are running on them. Specifically, a trusted VIM usually includes the following services: (1) secure boot process of both the controller and the compute nodes; (2) authentication process initiated by the

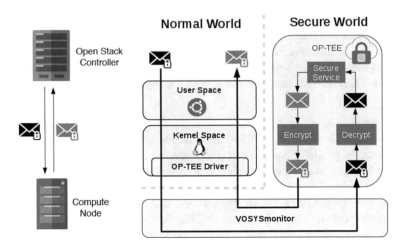

Fig. 2 Trusted VIM with OpenStack [53]

controller node towards the compute nodes for avoiding man-in-the-middle attacks; and (3) kernel integrity check for both the controller node and the compute nodes to avoid the corruption of the kernel memory. An example of an integration of a trusted OP-TEE environment in the OpenStack VIM platform is illustrated in Figure 2.

TEEs are also used for data protection in 5G multi-domain environments. Specifically, the use of NFV technologies enables resource sharing in multi-tenant network slices, which may also lead to eavesdropping by unauthorized or malicious entities. A consequence of such a cyber-attack would be data leakage across tenant slices or even more severe resource damage (e.g., in the 5G Core Network [48]) that would tear down the 5G infrastructure. Furthermore, the use of blockchain mechanisms, such as DLT technologies, makes the 5G multi-domain environment more vulnerable to lateral movements that would propagate the attack across different DLT nodes. The application of a TEE would allow maintaining the transparency of the DLT, whilst ensuring the privacy of data and applications [18]. This is accomplished using the ECC module (see Fig. 1) that encrypts the data associated with each transaction and stored locally in each DLT node. Apart from the data, keeping the bids secret is also of primary importance, so that neither another DLT node nor any other party can learn anything about them. Hence, encryption is also added to the transaction bids and the key to decrypt them resides only inside the trusted environment. In this way whenever each DLT node commits a bid in the blockchain, it is always encrypted and only trusted recipients that have the key can decrypt the bid to visualize the underlying transaction. Trust can be established using dedicated authentication or authorization methods. Encryption of the transaction bids ensures both the privacy and trustworthiness of data being exchanged between different DLT nodes.

Based on a literature review of the current trends and implementations, the following TEE mechanisms are feasible in 5G multi-domain environments, such as the one provided by 5GZORRO:

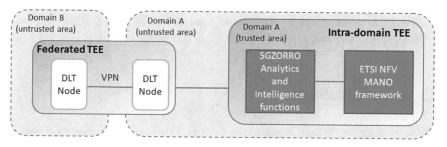

Fig. 3 TEE mechanisms for 5G multi-domain environments

1. *Intra-domain TEE*, in which the functional modules that are related to communication inside an administrative resource domain are executed securely through the presence of a software- or hardware-based TEE. Additionally, such TEE allows the function modules to exchange data in a protected manner. Each domain also includes an untrusted area, which has modules that are not executed inside a TEE component. The interactions between the intra-domain TEE and the untrusted area of a domain take place through secure internal channels to avoid a potential compromise of the intra-domain TEE.

2. *Federated TEE*, which focuses on the protection of the communication between different administrative resource domains. Such TEE could extend tunneling mechanisms for the protection of data exchange as well as to prevent eavesdropping from malicious entities.

These mechanisms are illustrated with an example within the scope of 5GZORRO in Fig. 3. Specifically, the Intra-domain TEE includes Analytics and Intelligence functions that are used for the automation of complex resource management procedures, such as the proactive scaling mechanism to increase or decrease the mobile infrastructure capacity through third-party resources based on tenant or user demands. These functions usually reside in the same administrative domain with the ETSI NFV MANO framework and inform it of any updates on the existing 5G network slices. On the other hand, DLT nodes of a blockchain environment are distributed across different domains offered services and hence are using Federated TEE mechanisms to communicate, such as the VPN mechanism that is shown in Fig. 3. The presence of a VPN ensures that the data remains protected when they are transmitted from one domain to another.

Even if the TEE mechanisms are already used for increasing the cyber-resilience in various application domains, their implementation on 5G multi-domain environments faces challenges as it does not cover the entire mobile network infrastructure. This is due to the hesitant nature of Mobile Network Operators (MNOs) towards changes that would interrupt the normal operation of their network and the services offered to their customers. Furthermore, if these changes occasionally involve the presence of new hardware as with hardware-based TEEs this hesitation becomes even greater. Moreover, even with the presence of TEEs a 5G infrastructure and the multi-domain environment is not adequately secure as additional challenges are also involved. These challenges are presented in the following section.

4.2 Future Challenges

The wide variety of 5G scenarios, where TEE may be applied, brings with it numerous challenges to be fulfilled through enriching current solutions with advanced technological approaches and features. In particular, the next challenges are principally centered on potential threats associated with TEE solutions, the impediment of employing TEE solutions in certain environments or contexts due to the performance, and finally, the agnosticism required by TEE solutions to be deployed or instantiated in multiple 5G enablers.

- **Pivotal threats and vulnerabilities**. One of the most significant challenges of TEE solutions is the threats related to microarchitectural attacks. Concretely, some of these paramount threats arise since TEE solutions intend to maximize performance gains. That is the case of side-channel attacks, one of the most well-known as identified in [39]. By means of this attack, an adversary may acquire user sensitive meta-information since it intends to discover the memory access patterns, and then, these patterns are exploited to infer target information. In fact, outstanding solutions such as Intel SGX [26] and Arm TrustZone [60] are not completely exempt from this attack. Another reiterated attack on TEEs is the transient execution [57], in which an adversary exploits the out-of-order execution of CPUs to leak actual data. In contrast to the side-channel attack, the transient execution attack is mostly applied to Intel SGX solutions [52]. Finally, two further cyberattacks are related to fake secure enclaves [31] and fault attacks [52], which are both associated with Intel SGX approaches. The first attack occurs when enclaves are not able to withstand memory corruption errors, and therefore, an attacker may partially execute arbitrary code inside the enclave. The second one is linked to tampering computations for deflecting data or control flow, as well as to the disruption of code or data.
- **Performance**. Due to the additional operations required to securely execute the tasks in the TEEs, the performance of the execution can be considerably downgraded. The impact is conditioned by the number and complexity of additional processing steps inside the TEE, so minimizing them without losing security properties is a key challenge in TEE applicability. Besides, TEE usage can require specific configuration and optimization, so it is critical to facilitate these steps. In this sense, Akram et al. analyzed in [14] how high-performance computing related to scientific tasks is affected when running on hardware-based TEEs (Intel SGX and AMD SEV). They showed how execution performance is considerably degraded when using default configurations. However, a proper configuration can secure the computing tasks without significant performance degradation, mainly in AMD SEV.

- **Agnostic solutions**. Since the use of TEEs ranges from embedded sensors to cloud servers, while considering a range of security risks [56], power constraints, and cost choices, it is crucial to abstract the complexity of TEE designs from their integration in dynamic environments. In this regard, vendors should analyze the feasibility of hardware-based, software-based, and integrated TEEs to facilitate their deployment and/or integration in multiple 5G enablers such as SDNs, NFVs, and DLTs. Thus, an organization named GlobalPlatform [2] is working on developing and publishing standards for TEE-related interfaces and implementations. At the software level, there are solutions such as OP-TEE [6] (using the TrustZone technology) that has support for the GlobalPlatform TEE Client API. In addition, this approach is being addressed in several areas of current literature, to name a few, the trusted VIM in the cloud and edge environments [53] or IoT devices [36]. Nonetheless, while there are many hardware solutions that can support a TEE, there are not enough agnostic approaches to apply a software-based TEE solution to multiple hardware-based TEE solutions.

Inasmuch as the above-mentioned trends and challenges, it is possible to gather a set of research lines that should be considered for future developments of TEE solutions. In that sense, the forthcoming investigations will mainly focus on the TEE consideration in 5G multi-domain and multi-stakeholder environments, as well as its integration with existing NFV MANO architectures and their crucial components such as VIMs and NFVIs.

5   Conclusions and Future Work

The present chapter has shown a proposal for integrating the use of a hardware-based TEE solution, like SCONE, with 5G technologies, which enables the deployment of network components in third-party infrastructures. The proposed TEEaaS framework attempts not only to cover security and trust aspects under 5G core network components such as VIMs or VNFs, but also to consider pivotal performance and delay requirements. Thus, the proposed boosts the trustworthy execution of offloading critical data and processes on demand. In particular, the TEEaaS framework is developed considering the 5GZORRO H2020 project requirements and properties, where the trading of heterogeneous resources among stakeholders is contemplated to make easier the establishment of thoroughly pervasive services across different domains. Furthermore, a set of current trends and future challenges has been recognized during the state-of-the-art review and design phase, which will be considered during the development phase.

As future work, we plan to finish the implementation of the TEEaaS framework, since it should enable its deployment and integration with other 5GZORRO resources such as SLA monitoring. Additionally, the framework design will be aligned with the trends and challenges that have been discussed in this book chapter. To validate the suitability of the proposed solution, we plan to carry out a Proof-of-Concept (PoC) which will protect not only data and software while running on the secure enclave, but also while data is in transit and at rest. Finally, we plan to use the PoC to measure the resilience of well-known threats and vulnerabilities, and on the other hand, the fulfillment of performance requirements in 5G multi-domain networks.

Acknowledgements This work has been supported by the European Commission through 5GZORRO project (grant no. 871533) part of the 5G PPP in Horizon 2020. The paper solely reflects the views of the authors. EC is not responsible for the contents of this paper or any use made thereof. Authors thank the 5GZORRO Consortium for useful insights to this work.

References

1. Genode—An exploration of Arm TrustZone technology. https://genode.org/documentation/articles/trustzone. Accessed 04 May 2021
2. GlobalPlatform. https://globalplatform.org/. Accessed 04 May 2021
3. GlobalPlatform device technology. TEE client API specification. https://globalplatform.org/specs-library/tee-client-api-specification/. Accessed 04 May 2021
4. Intel Trusted Execution Technology (Intel TXT) overview. https://www.intel.com/content/www/us/en/support/articles/000025873/technologies.html. Accessed 04 May 2021
5. Multizone security for RISC-V. https://hex-five.com/multizone-security-sdk/. Accessed 04 May 2021
6. Open Portable Trusted Execution Environment (OP-TEE). https://github.com/OP-TEE. Accessed 04 May 2021
7. SDN 5G threat analysis: An ENISA case study. https://www.charisma5g.eu/wp-content/uploads/2016/07/SDN5G-Threat-Analysis-An-ENISA-case-study.pdf. Accessed 04 May 2021
8. Secure Enclaves for Reactive Cloud Applications (SERECA) project. https://www.serecaproject.eu. Accessed 04 May 2021
9. Secure Linux Containers (SCONE). https://sconedocs.github.io/aboutScone/. Accessed 04 May 2021
10. SecureCloud. https://www.securecloudproject.eu/. Accessed 04 May 2021
11. Solacia SecuriTEE. https://www.insidesecure.com/Company/Press-releases/Inside-Secure-AND-SOLACIA. Accessed 04 May 2021
12. Trusted Executed Environment (TEE), Trustonic. https://www.trustonic.com/technical-articles/what-is-a-trusted-execution-environment-tee/. Accessed 04 May 2021
13. Trusted Little Kernel. https://trustedfirmware-a.readthedocs.io/en/latest/components/spd/tlk-dispatcher.html. Accessed 04 May 2021
14. Akram, A.: Performance analysis of scientific computing workloads on general purpose TEEs. In: 35th IEEE International Parallel & Distributed Processing Symposium (IPDPS). IEEE (2021)
15. Alrawais, A., Alhothaily, A., Hu, C., Xing, X., Cheng, X.: An attribute-based encryption scheme to secure fog communications. IEEE Access **5**, 9131–9138 (2017)
16. Amacher, J., Schiavoni, V.: On the performance of arm trustzone. In: IFIP International Conference on Distributed Applications and Interoperable Systems, pp. 133–151. Springer (2019)
17. Arfaoui, G., Bisson, P., Blom, R., Borgaonkar, R., Englund, H., Félix, E., Klaedtke, F., Nakarmi, P.K., Näslund, M., O'Hanlon, P., et al.: A security architecture for 5G networks. IEEE Access **6**, 22466–22479 (2018)
18. Ayoade, G., Karande, V., Khan, L., Hamlen, K.: Decentralized IoT data management using blockchain and trusted execution environment. In: 2018 IEEE International Conference on Information Reuse and Integration (IRI), pp. 15–22. IEEE (2018)

19. Baldoni, G., Cruschelli, P., Paolino, M., Meixner, C.C., Albanese, A., Papageorgiou, A., Khalili, H., Siddiqui, S., Simeonidou, D.: Edge computing enhancements in an NFV-based ecosystem for 5G neutral hosts. In: 2018 IEEE Conference on Network Function Virtualization and Software Defined Networks (NFV-SDN), pp. 1–5. IEEE (2018)
20. Bangerter, B., Talwar, S., Arefi, R., Stewart, K.: Networks and devices for the 5G era. IEEE Commun. Mag. **52**(2), 90–96 (2014)
21. Barros Lourenço, M., Marinos, L., Patseas, L.: ENISA Threat Landscape for 5G Networks. Tech. rep, European Union Agency for Cybersecurity (2020)
22. Bendriss, J., Yahia, I.G.B., Chemouil, P., Zeghlache, D.: AI for SLA management in programmable networks. In: DRCN 2017-Design of Reliable Communication Networks; 13th International Conference, pp. 1–8. VDE (2017)
23. Bhushan, K., Gupta, B.B.: Distributed denial of service (DDoS) attack mitigation in software defined network (SDN)-based cloud computing environment. J. Ambient Intell. Hum. Comput. **10**(5), 1985–1997 (2019)
24. Bondan, L., Franco, M.F., Marcuzzo, L., Venancio, G., Santos, R.L., Pfitscher, R.J., Scheid, E.J., Stiller, B., De Turck, F., Duarte, E.P., et al.: FENDE: marketplace-based distribution, execution, and life cycle management of VNFs. IEEE Commun. Mag. **57**(1), 13–19 (2019)
25. Boos, P., Lacoste, M.: Networks of trusted execution environments for data protection in cooperative vehicular systems. In: Vehicular Ad-hoc Networks for Smart Cities, pp. 99–109. Springer (2020)
26. Brasser, F., Müller, U., Dmitrienko, A., Kostiainen, K., Capkun, S., Sadeghi, A.R.: Software grand exposure: SGX cache attacks are practical. In: 11th USENIX Workshop on Offensive Technologies (WOOT) (2017)
27. Carrozzo, G., Siddiqui, M.S., Betzler, A., Bonnet, J., Perez, G.M., Ramos, A., Subramanya, T.: AI-driven zero-touch operations, security and trust in multi-operator 5G networks: a conceptual architecture. In: 2020 European Conference on Networks and Communications (EuCNC), pp. 254–258. IEEE (2020)
28. Celdran, A.H., Perez, M.G., Clemente, F.J.G., Perez, G.M.: Enabling highly dynamic mobile scenarios with software defined networking. IEEE Commun. Mag. **55**(4), 108–113 (2017)
29. Celdrán, A.H., Pérez, M.G., Clemente, F.J.G., Pérez, G.M.: Towards the autonomous provision of self-protection capabilities in 5G networks. J. Ambient Intell. Hum. Comput. **10**(12), 4707–4720 (2019)
30. Chhabra, M., Gupta, B., Almomani, A.: A Novel Solution to Handle DDOS Attack in MANET (2013)
31. Cloosters, T., Rodler, M., Davi, L.: TeeRex: discovery and exploitation of memory corruption vulnerabilities in SGX enclaves. In: 29th USENIX Security Symposium (USENIX Security 20), pp. 841–858 (2020)
32. Costan, V., Devadas, S.: Intel SGX explained. Cryptology ePrint Archive, Report 2016/086 (2016). https://eprint.iacr.org/2016/086
33. Elbashir, K.: Trusted execution environments for open vswitch: a security enabler for the 5G mobile network (2017)
34. Elgendy, I.A., Zhang, W., Tian, Y.C., Li, K.: Resource allocation and computation offloading with data security for mobile edge computing. Future Gener. Comput. Syst. **100**, 531–541 (2019)
35. Esposito, C., Ficco, M., Gupta, B.B.: Blockchain-based authentication and authorization for smart city applications. Inf. Process. Manag. **58**(2), 102468 (2021)
36. Göttel, C., Felber, P., Schiavoni, V.: Developing secure services for IoT with OP-TEE: a first look at performance and usability. In: IFIP International Conference on Distributed Applications and Interoperable Systems, pp. 170–178. Springer (2019)
37. Guerzoni, R., Vaishnavi, I., Perez Caparros, D., Galis, A., Tusa, F., Monti, P., Sganbelluri, A., Biczók, G., Sonkoly, B., Toka, L., et al.: Analysis of end-to-end multi-domain management and orchestration frameworks for software defined infrastructures: an architectural survey. Trans. Emerg. Telecommun. Technol. **28**(4), e3103 (2017)

38. Gupta, B., Quamara, M.: An overview of internet of things (IoT): architectural aspects, challenges, and protocols. Concurrency Comput.: Pract. Exp. **32**(21), e4946 (2020)
39. Jauernig, P., Sadeghi, A.R., Stapf, E.: Trusted execution environments: properties, applications, and challenges. IEEE Secur. Priv. **18**(2), 56–60 (2020)
40. Jiang, F., Fu, Y., Gupta, B.B., Liang, Y., Rho, S., Lou, F., Meng, F., Tian, Z.: Deep learning based multi-channel intelligent attack detection for data security. IEEE Trans. Sustain. Comput. **5**(2), 204–212 (2020). https://doi.org/10.1109/TSUSC.2018.2793284
41. Kim, S., Han, J., Ha, J., Kim, T., Han, D.: Enhancing security and privacy of Tor's ecosystem by using trusted execution environments. In: 14th USENIX Symposium on Networked Systems Design and Implementation (NSDI), pp. 145–161 (2017)
42. Lefebvre, V., Santinelli, G., Müller, T., Götzfried, J.: Universal trusted execution environments for securing SDN/NFV operations. In: 13th International Conference on Availability, Reliability and Security, pp. 1–9 (2018)
43. Liu, Y., An, K., Tilevich, E.: RT-Trust: Automated refactoring for different trusted execution environments under real-time constraints. J. Comput. Lang. **56**, 100939 (2020)
44. Mijumbi, R., Serrat, J., Gorricho, J.L., Latré, S., Charalambides, M., Lopez, D.: Management and orchestration challenges in network functions virtualization. IEEE Commun. Mag. **54**(1), 98–105 (2016)
45. Nour, B., Ksentini, A., Herbaut, N., Frangoudis, P.A., Moungla, H.: A blockchain-based network slice broker for 5G services. IEEE Netw. Lett. **1**(3), 99–102 (2019)
46. Ordonez-Lucena, J., Ameigeiras, P., Lopez, D., Ramos-Munoz, J.J., Lorca, J., Folgueira, J.: Network slicing for 5G with SDN/NFV: concepts, architectures, and challenges. IEEE Commun. Mag. **55**(5), 80–87 (2017)
47. Ortiz, J., Sanchez-Iborra, R., Bernabe, J.B., Skarmeta, A., Benzaid, C., Taleb, T., Alemany, P., Muñoz, R., Vilalta, R., Gaber, C., et al.: INSPIRE-5Gplus: intelligent security and pervasive trust for 5G and beyond networks. In: 15th International Conference on Availability, Reliability and Security, pp. 1–10 (2020)
48. Park, S., Choi, B., Park, Y., Kim, D., Jeong, E., Yim, K.: Vestiges of past generation: threats to 5G core network. In: International Conference on Innovative Mobile and Internet Services in Ubiquitous Computing, pp. 468–480. Springer (2020)
49. Sabt, M., Achemlal, M., Bouabdallah, A.: Trusted execution environment: what it is, and what it is not. In: 2015 IEEE Trustcom/BigDataSE/ISPA, vol. 1, pp. 57–64. IEEE (2015)
50. Sangorrín, D., Honda, S., Takada, H.: Reliable and efficient dual-OS communications for real-time embedded virtualization. Inf. Media Technol. **8**(1), 1–17 (2013)
51. Santos, N., Raj, H., Saroiu, S., Wolman, A.: Using Arm TrustZone to build a trusted language runtime for mobile applications. In: 19th International Conference on Architectural Support for Programming Languages and Operating Systems, pp. 67–80 (2014)
52. Schwarz, M., Gruss, D.: How trusted execution environments fuel research on microarchitectural attacks. IEEE Secur. Priv. **18**(5), 18–27 (2020)
53. Sechkova, T., Barberis, E., Paolino, M.: Cloud & edge trusted virtualized infrastructure manager (VIM)-security and trust in OpenStack. In: 2019 IEEE Wireless Communications and Networking Conference Workshop (WCNCW), pp. 1–6. IEEE (2019)
54. Stergiou, C.L., Psannis, K.E., Gupta, B.B.: IoT-based big data secure management in the fog over a 6G wireless network. IEEE Internet Things J. (2020)
55. Swapna, A.I., Rosa, R.V., Rothenberg, C.E., Sakellariou, I., Mamatas, L., Papadimitriou, P.: Towards a marketplace for multi-domain cloud network slicing: use cases. In: 2019 ACM/IEEE Symposium on Architectures for Networking and Communications Systems (ANCS), pp. 1–4. IEEE (2019)
56. Tewari, A., Gupta, B.: Security, privacy and trust of different layers in internet-of-things (IoTs) framework. Future Gener. Comput. Syst. **108**, 909–920 (2020)
57. Van Bulck, J., Minkin, M., Weisse, O., Genkin, D., Kasikci, B., Piessens, F., Silberstein, M., Wenisch, T.F., Yarom, Y., Strackx, R.: Foreshadow: extracting the keys to the intel SGX kingdom with transient out-of-order execution. In: 27th USENIX Security Symposium (USENIX Security 18), pp. 991–1008 (2018)

58. Xiao, Y., Zhang, N., Li, J., Lou, W., Hou, Y.T.: PrivacyGuard: Enforcing private data usage control with blockchain and attested off-chain contract execution. In: European Symposium on Research in Computer Security, pp. 610–629. Springer (2020)
59. Xu, S., Qian, Y., Hu, R.Q.: Privacy-preserving data preprocessing for fog computing in 5G network security. In: 2018 IEEE Global Communications Conference (GLOBECOM), pp. 1–6. IEEE (2018)
60. Zhang, N., Sun, K., Shands, D., Lou, W., Hou, Y.T.: TruSpy: Cache side-channel information leakage from the secure world on Arm devices. Cryptology ePrint Archive, Report 2016/980 (2016). https://eprint.iacr.org/2016/980
61. Zhu, J., Hou, R., Wang, X., Wang, W., Cao, J., Zhao, B., Wang, Z., Zhang, Y., Ying, J., Zhang, L., Meng, D.: Enabling rack-scale confidential computing using heterogeneous trusted execution environment. In: 2020 IEEE Symposium on Security and Privacy (SP), pp. 1450–1465 (2020). https://doi.org/10.1109/SP40000.2020.00054

WSNs and IoTs for the Identification of COVID-19 Related Healthcare Issues: A Survey on Contributions, Challenges and Evolution

Anish Khan and B. B. Gupta

Abstract In this era of technological revolution, we are familiar with many scientific terminologies and gadgets. Today, internet is the backbone of the whole world as internet connectivity plays a vital role in our routine life and makes our life much easier. Wireless sensors are implemented in many applications like agricultural sector, military, home automation and health care sector. These wireless sensors are easy to operate and handle. Their performance varies according to the application. By connecting internet with these smart wireless sensors they act like Internet of Things. In the present scenario, whole world is suffering from Covid-19 pandemic. This is very strenuous situation for mankind. It is enigmatic to recognize a person with Covid-19 symptoms. For the identification of affected patient, some models are coined with the aid of wireless sensors and internet of things. The principle goal of this survey is to demonstrate the critical role of wireless sensor networks with internet of things for Covid-19 health care purposes.

Keywords Wireless sensor network (WSN) · Internet of things (IoT) · Sensor nodes (SN) · COVID-19 · Corona-virus · World Health Organisation (WHO)

1 Introduction

Wireless sensor network is defined as the network of wireless remote devices that will gather, process and forward data to the controlling station [1–3]. Most recent innovations offer us affordable and little programmed gadgets known as "Sensor Nodes" to screen the physical and ecological conditions, for example, pressure, temperature etc. These nodes collect data from remote locations. Nodes havean onboard battery that will define the lifespan of node [4]. A network comprises various nodes that will capable of communicating to a specific range. The sensor nodes consist

A. Khan (✉)
UIET, Kurukshetra University, Kurukshetra, India

B. B. Gupta
National Institute of Technology, Kurushatra Haryana, India

© The Author(s), under exclusive license to Springer Nature Switzerland AG 2022
A. A. Abd El-Latif et al. (eds.), *Security and Privacy Preserving for IoT and 5G Networks*,
Studies in Big Data 95, https://doi.org/10.1007/978-3-030-85428-7_10

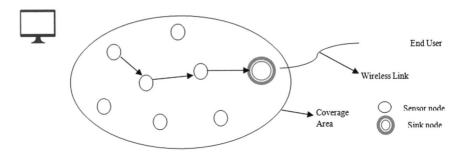

Fig. 1 Architecture of WSN

of onboard equipments like onboard processor, memory storage, battery etc [5–9]. The nodes continuously sense their surroundingsand sensed data is collaborated for further processing. These devices have restricted battery lifetime. Analysis of power is important for better efficiency. The requirement for energy efficient arrangement for WSN is becoming progressively important.

As we can see the Fig. 1, that gives us a bird's eye view of what a network looks like. When a sensor node assembles data from the surroundings, it transfers to the BS, which is the managing centre of the WSN [10, 11]. The transferring of messages in network is an energy consuming exercisethat affects network lifetime. WSN's are categorised in two categories: homogenous andheterogeneous networks. In the Homogenous network, all sensors are same in relationtoenergy & hardware. In a Heterogeneous network, diverse kinds of sensor node have different battery power [12–15]. Compact size, less power consumption, on-board computational capabilities of sensors are the results of the advancements in Micro Electro Mechanical Systems (MEMS) [16].The sensor nodes are haphazardly positionedwithin or near the sensed observable fact.The batteries of the sensor can't be recharged or replaced thus the utilization of energy is the primary system configuration issue in WSN.

1.1   Types of Applications

1.1.1   Event Detection

It is clear that, in a scenario there are source nodes that are used to sense the data and other is sink node where data is to be transferred. In the event detection application, whenever sensor senses the data or a happening event, they immediately forward sensed information to the sink node [17–20]. For example, a change in temperature weather the temperature is raised or decreased a set limit it forwards the information to the sink node.

1.1.2 Periodic Measurement

There are many applications where periodic measurements are required, for example, in a nuclear chamber; there is a risk of raise in temperature and leakage of the chemicals or harmful gases. To monitor these kinds of information that are very dangerous to the mankind, periodic measurements are required [21–23].

1.1.3 Tracking

The nature of an event can be mobile. In most of the cases of surveillance, a wireless sensor network is used to track the event with the position, direction and speed. The report is continuously updated at the base station [24–29]. For example, some tracking sensors are available these days that are placed inside the car or any vehicle. Whenever the car or motorbike is far away from you, tracking sensor helps us to find out our vehicle. It prevents vehicle from thieves.

1.1.4 Deployment Options

Wireless sensor network provides a wide range of deployment options. As the sensor nodes can have a fixed architecture or it can be deployed randomly from air in the forest or in any tropical structures [30–33]. Moreover sensor nodes can have their own mobility so that they can easily move from one place to another.

1.1.5 Maintenance Options

We know that sensors are usually placed out of human reach or they are deployed where it is not easy to reach out there. Hence the maintenance is bit difficult task for the nodes. The main issue is their battery life [34–36]. The drainage of battery is more frequent so it is not possible to change or replace the battery frequently (Fig. 2).

1.2 Applications of Wireless Sensor Network

1.2.1 Military Applications

WSNis an integral part of military communication intelligence and targeting systems. In the military, the wireless sensor networks are used for security purposes. The nodes are deployedunderthe area of observation.The application example of sensors is to detect enemy intrusion. In security monitoring there is no aggregation of the datainstead the nodes frequently check the status and transmit the report when there is a security violation [37–42].

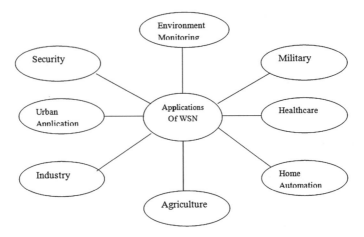

Fig. 2 Applications of wireless sensor network

1.2.2 Environmental Monitoring

This sensor measures light, temperature, and humidity, and can be equipped to do soil-moisture measurements. WSNs are also used as process monitoring, forestry fire recognition, landslide industrial management. Nowadays,there are many types of sensors available in the market, like temperature sensors, pressure sensors, light sensors or any other sensors to detect and monitor various phenomena related real-world WSN applications [43–47].

(a) **Disaster relief applications**:
It is regularly referenced application of WSN for disaster relief activities. We have a most common example of wildfire detection; in this, nodes are integrated with temperature sensors on them and are identified by their location coordinates in the scenario. Nodes are placed randomly from air with the help of aeroplanes like in wildfire (forests). After deployment, nodes generate a temperature graph for the particular area and that graph is interpreted outside from the spot by fire-fighters. This same scenario can be implemented inside the chemical chambers to avoid any disaster [48–51].

1.2.3 Urban Applications

(a) **Intelligent buildings**:
In giant buildings, there are many aspects that are to be taken under observation like, humidity level, improper ventilation, air conditioning etc also known as "HVAC". Due to these aspects, a large amount of energy gets devoured by buildings [52]. To provide a better or real time monitoring of air flow, temperature and many other parameters inside a giant structure, WSN improves comfort and

ease of life with less energy consumption. The giant structure like "Burj-Khalifa" in Dubai is one of the good examples of "Intelligent building". Moreover sensors are also used to calculate the mechanical stress level and bending load of girders of the buildings. So that it can minimise the cause of any miss happening in the future. The same scenario can be utilized over bridges [53–58].

1.2.4 Health Applications

The WSN's are used for monitoring the movements and internal processes of organs etc. The informationgathered by nodes can be kept for long period, and information is used for medical analysis. The sensorsare also used for tracking and monitoring of patients. To analyze patient health, small sensors are attached on their body so that doctors have live reports of their patients [59–64]. Many sensors are available in market to that will perform specific tasks. For example, sensors are usedto detect heart beat andpulse rate, another is to detect blood pressure of patient.

1.2.5 Agriculture Application

In the field of agriculture, sensors are used to measure the temperature in the field, moisture present in the soil and humidity of surroundings. All these physical parameters are detected by the sensors that are deployed inside the field. This will help farmers to make their farming effective [65, 66].

1.2.6 Security

(a) **Facility management**:
    WSNhasan extensive variety of applications which facilitate the inhabitants in various forms. Let us take an example of a "key less entry" application. This application is very popular in multi-national companies, where authentication is required to enter the company area or to access their private property [67]. In this application the people who are caring their badges are allowed to enter/access the things. The badges are scanned by sensors which allow or deny people to access their private things. Moreover, another application where WSN is used frequently is in parking area, where vehicles having tags are allowed to enter in parking slots. WSN is also used to detect vehicles equipped with a tracking sensor, which is installed inside the vehicle and provides information about the vehicle's location. Another application is, WSN can be deployed inside or near around the nuclear chambers to track the temperature and any type of leakage [68].

(b) **Logistics**:
    As there are numerous applications of the wireless sensor network, in the field of logistics, WSN are used commonly. As there is a huge amount of import and

exportstaking place across whole world and in many countries. Not only outside the country, there are too many items that are being supplied inside the country for households. WSN is used to identify such items by using the RFID tag "Radio Frequency Identifier" [69–72]. However, when we are at the airport, these tags are used to identify our luggage on the conveyor belts.

1.2.7   Industry

(a) **Machine surveillance**:
   The deployment of the sensor nodes is done where it is not easy or sometimes impossible to reach. As it is a period of modern era, technology is at its peak. We have heavy and more complex structure of machines present these days. When these machines are facing some trouble, sometimes it is not possible to detect the problems that are faced by the machinery. So, we placed sensors in the complex structure that give us the proper track record. If any problem occurs inside the machine, then it is easy to detect and repair the machinery [73–75].

1.3   Sensor Node Architecture

As we all know it is an era of technology, many small size sensors are fabricated these days. They are capable of making computations and they are very intelligent too. Wireless sensor network basically consists of fourmain units that are power supply, memory, CPU, communication system [76–81]. They all have their own capabilities and these all components combined together to make a sensor. The power supply provides battery backup so that the node will stay alive for a long period, CPU is used for making computation on the data sensed, memory is used to store the sensed data, and a communication system is used to transfer the information to sink node or base station. The typeof sensor being used in a sensor node will depend on the application.

A basic sensor node comprises of five main components as shown in Fig. 3.

1.3.1   Controller

Controller is the fundamental element of a sensor node. The main function of the controller is to accumulate information from other sensors. Following data collection, the controller applies operations over the data and takes decision where and when to deliver the data. It is the CPU of sensor node. A number of operations are applied to data. Sensor nodes are making use of microcontrollers instead of general purpose processors that are used in desktop computers because general purpose processors consume more power [82–86]. But microcontrollers are suitable for embedded systems because their energy consumption is very low as compared to

Fig. 3 Sensor node
architecture

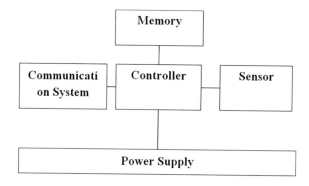

desktop processors, they have inbuilt memory, are easy to program hence microcontrollers are very flexible. Some of the common microcontrollers that are being used
in sensor nodes:

(i)   Intel StrongARM
(ii) Texas Instruments MSP 430
(iii) Atmel ATmega.

1.3.2   Memory

The memory is used to store the data that is sensed or received from nearby sensor
nodes. Random Access Memory (RAM) is used to store the information because;
RAM is fast in processing and stores sensor readings or information packets from
other sensors that are present inside the network immediately. RAM has only one
disadvantage that it will erase its data whenever power failure occurs. Read Only
Memory (ROM) is used to store the program code [87–89].

1.3.3   Sensors and Actuators

Without sensors & actuators, WSN would be irrelevant completely. As there are
numerous applications present nowadays. It is a difficult task to identify the sensors for the particular application [27, 51, 62]. The sensors are divided into three
categories:

(i)   Omnidirectional sensors
(ii) Narrow beam sensors
(iii) Active sensors.

Omnidirectional sensors are passive sensors that interpret physical quantities without
manipulating the environment. The main functionality of these sensors is that they are
self powered and they obtain energy from the environment. Direction is not important

in omnidirectional sensors [35, 73, 76]. The main examples of these sensors are thermometer, humidity, smoke detector, air pressure etc. Narrow beam sensors are also passive sensors. The main difference between omnidirectional and narrow beam sensors is they require direction of measurement. Most popular example is camera [10, 21, 47]. Active sensors are very intelligent type of sensors. They investigate the environment in a good manner. The main example of active sensors is, sonar radars use these sensors [6, 33, 84]. Moreover a sensor has its coverage limit, a sensor can sense in its coverage area.

1.3.4 Communication

Exchange of data is done with the help of communicating devices. It is totally depends upon the user what medium they want to choose for communication. As there are two kind of mediums for communication, one is wired and the other is wireless [31, 48, 74]. When we talk about the communication in the WSN, only wireless medium is used for communication. For the communication, transmitter and receiver are used at the sensor node which is part of architecture of node. The main objective of these two components is to convert bit stream that is coming from the controller into radio waves so that the information is being exchanged with other nodes. As we know, the size of the node is very small hence, we combined both components and the resultant component is known as transceiver [7, 20, 70, 76]. Here are some characteristics of transceiver that are used for the wireless communication as discussed below:

(i) Carrier frequency: It can also be defined as "channels". Transceivers are operated at various frequencies. For example MAC protocol CSMA/ALOHA techniques [29, 56].
(ii) Power consumption: It is defined as the power required to transmit or receive a single bit is called power consumption. There may be two states, like sleep state and awake state [38, 61, 83].
(iii) Modulation: Various modulation techniques are used like ASK, PSK etc [22, 53].
(iv) Noise figure: Itis the ratio of SNR at input and SNR at output [29, 38, 64].

$$NR = \frac{SNR_i}{SNR_o} \tag{1}$$

Transceiver structure
Figure 4 represents the transceiver operational modes [77, 87, 90, 91]:
Transmit: In transmit mode, transmitter is active and antenna starts radiating energy.
Receive: In receive mode, receiver is active.
Sleep: In sleep mode, transceiver is in sleep state, no transmission or reception is conducted by the transceiver. Despite the fact that IEEE 802.11 provides numerous sleep modes according to that transceiverwill function.

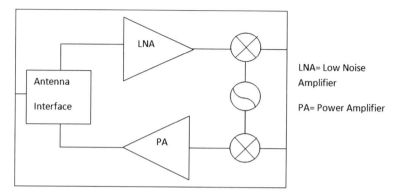

Fig. 4 Radio front end

Idle: In idle mode, transceiver is ready to receive but currently it is not receiving anything.

1.3.5   Power Supply

It is more important component as compared to other four components that are integrated on the sensor node. As whole network is depends on sensor nodes and the sensor nodes are dependent on power supply. We know very well the nodes are placed where there is impossible to reach or out of human reach hence the entirescenario is relianton batteries. It is not possible to replace or recharge the battery of nodes. Hence to make our network more reliable with an extended life span of the network we have to take energy consumption as the key player [91–95].

1.4   Challenges for WSN

1.4.1   Type of Service

Type of service is defined as the transferring of bits from one place to another place. The meaning of transferring bits illustrates, communicating useful or redundant data for a particular task. "People want answers, not numbers" said by Steven Glaser.

1.4.2   Quality of Service

It is defined by quality of network that provides services to the users. The major service requirements are less delay, minimum bandwidth consumption, low latency rate, minimum error and provides more reliability while transferring the multimedia data over a network [96–98].

1.4.3 Fault Tolerance

In a wireless network, whole scenario works on wireless communication. All the nodes that are present in the network will have their own energy life time that is directly dependent on battery they carefor them. As sensor deployment is ususlly random,it will leads in the distruction of some nodes while they are being placed. So when the communication begins in progress between two nodes and suddenly one of them or both of them will loss there battery, then communication failure will occurs. To overcome this failure, the WSN is capable of makingcommunication frequent with no fault occurrence [99–101].

1.4.4 Lifetime

As we know, the sensors are fully dependent on limited power supply usually a battery that is integrated on sensor node. To recharge or replace the battery is not an easy task in WSN because of their deployment strategies. Once the battery has fully discharged the node will die. Hence the lifetime of the network relays on the battery lifetime of the sensors. To avoid the drainage of power, we can integrate solar cells so that it may recharge the battery. The network's quality of service and lifetime are inversely proportional. If users want a high quality network then it may happen that the lifetime of the network will degrade quickly [2, 102, 103].

1.4.5 Wide Range of Densities

The density of a network is defined as the "number of sensor nodes per unit area". Density of the network depends upon the application. Different type of applications have different densities. However, density does not always remain the same, it will change when ever node dies [95, 104, 105].

2 Internet of Things

The "Internet" is a worldwide group of combined servers, PCs, tablets and mobiles that are administered by standard protocols for combined frameworks. This enables users to send, receive and communicate information [106].

The word "Things" has many meanings in English dictionary. The word thing refers to an object, action and situation. For example, a mobile is referred to an object, 'those kinds of things are expected from her'- here things are referred as action [107]. With the combination of aforementioned terms, a new term originates called "Internet of Things" which refers to a network of physical devices or embedded devices which are capable of sending and receiving information or data via internet know as "Internet of Things" [108]. Vision of Internet of Things is to make things (tube-light, AC, fan, door bell, table etc) smart and to act like living entities by using internet. The internet has become ubiquitous and spread almost every part of the world and human life is directly influenced by internet.

Moreover Internet of Things has no fixed definition; many definitions are available in the literature. Let's have a brief view of some definitions that are available in the literature, Mark-Weiser gave a central statement in seminal paper "The most profound technologies are those that disappear. They weave themselves into the fabric of everyday life until they are indistinguishable from it" at Scientific American in 1991 [109]. An expert on digital innovation named Kevin Ashton stated "An open and comprehensive network of intelligent objects that have the capacity to auto-organize, share information, data and resources, reacting and acting in face of situations and changes in the environment" in 1998 [110].

Internet of Things (IOT) becomes a buzzword nowadays, it builds a communication bridge between physical world and cyber world. However, IOT is not an independent technology.It came into existence with the agglomeration of various technologies that works cooperatively to execute complex tasks with intelligence. IOT devices are heterogeneous in nature [111].They all have different capabilities in terms of sensing range, processor computation, memory, battery capacity [112–115]. With the passage of time much advancement is being done in the field of embedded systems and nano-electronics devices that makes possible the reduction of size and dramatic improvements in computational processes. These nano-electro devices are now available with global positioning systems (GPS), attached with things to make them visible and provide a real time monitoring system [116–120].

2.1 Internet of Things: Architecture

As mentioned above, IOT is not an individual technology; it is a cluster of different technologies. Hence, to implement this technology, various IOT architectures are available in literature proposed by different researchers. Some of the IOT architectures are discussed below [121–124] (Fig. 5).

2.1.1 Three-Layer Architecture

This is the basic architecture that was proposed in early research stages by researches, as when this technology newly came in existence it connects number of devices to the internet that makes much traffic over network and need for more storage increased. To overcome these issues, a three layer architecture has been developed. It uses three different layers that perform their own task [125–130]. The layers are Perception layer, Network Layer, Application layer as showm in Fig. 6.

Perception Layer: This is also known as "Device Layer", which comprise of sensors that are responsible for the sensing and collecting data from sensors. Sensors can

Fig. 5 Generic scenario of IOT

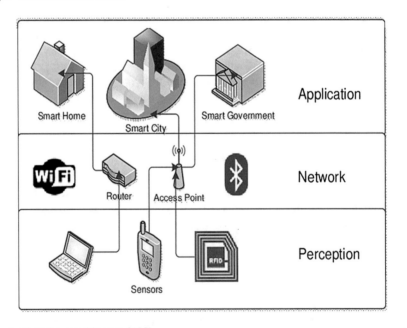

Fig. 6 Three layer architecture [125]

be Radio Frequency Identification tag(RFID tags), Barcodes, Temperature sensors, humidity sensors, Near Field Communication sensors etc. After gathering information from sensors, it is forwarded to the next layer i.e. Network Layer [125–131].

Network Layer: This layer can be referred to "Transmission Layer". The main purpose of this layer is to transfer useful information securely by means of wired or wireless communication to the upper layer i.e. Application Layer [125–130, 132, 133]. Moreover communication technology can be Bluetooth, ZigBee, WiFi etc.

Application Layer: The topmost layer of the three layered architecture is Application Layer. The main functioning purpose of this layer is to provide application specific service to the user [125–130, 134]. Various applications are offered by IOT technology for example, Health care applications, Smart home, Smart City, Smart car etc.

2.1.2 Five Layered Architecture

Three layer architecture (Fig. 6) describes the principle thought of the IoT, however it isn't adequate for research on IOT. Improvement in IOT relies on numerous new applications to improve the architecture a five layered architecture was designed. It consists of two new layers i.e. Middleware Layer, Business Layer [125–128, 130, 135–137]. However the three layers in previous model remain the same and their work is also same. Here we discussthe two newly introduced layers as illustrate in Fig. 7:

Middleware Layer: The main function of this layer is to provide service management. It has an uninterrupted link to the database. Information details are provided by the Network Layer, after receiving information it carries out some computations over the received information [125–128, 130, 135–137].

Business Layer: The overall management of the IOT system comprising the application and service is supervised by this layer. Plotting of graphs, flow charts and business models are done on the basis of provided data from the application layer [125–128, 130, 135–137].

In addition to this, there are many more network architectures are proposed with the state-of-the-art for example, Open System Interconnection (OSI) Model, Transmission & Control Internet Protocol TCP/IP, Fog Architecture, Telecommunication Management Network (TMN) Logical Layered Architecture etc.

2.2 Technologies of IOT

The process of identification of objects and to abolish human interference, "electronic tags" is used. The characteristic of electronic tags is that they can be readable at a small distance with the help of an electronic tag reader and can store information

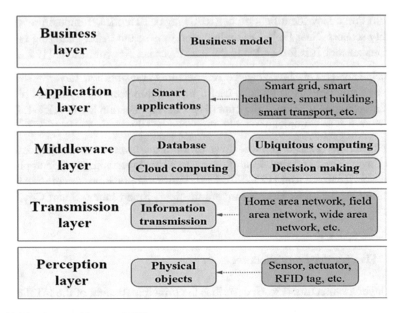

Fig. 7 Five-layer architecture [130]

about the physical device. Some of electronic tags are RFID tag, EPC tag, Barcode etc [114, 118, 131, 138–143]. In addition to this many sensors and actuators are used to collect the information of the smart devices like temperature sensors, ECG health care sensors, IR sensors, humidity sensors, vibration sensors [114, 118, 131, 138–143] etc. These sensors make improvements in the field of IOT technology. Here we discuss some of the technologies:

(i) **Electronic Product Code (EPC)**: EPC is defined as a unique identifier (UID) to every object or physical device that is present on the globe. This is supervised by EPC Global. EPC has a 96 bit length that consist of the following domain: Header (8 bits)

- EPC Manager (28 bits)
- Object Class (24 bits)
- Object Identification Number (36 bits).

EPC doesn't share anything about the characteristics of the product/device. Suppose if two products share the same properties but their manufacturers are different, then the devices will have different EPC's [114, 118, 131, 138–143]].

(ii) **RFID Tags**: It stands for Radio Frequency Identification Tag that comprises the EPC of respective physical device. RFID is classified into two main categories:

- **Active Tag**: active tag supports on-board battery supply. Their life span is limited and directly proportional to the life span of the battery attached.

They can stain a year of services due to the advancement in the nano-electonics devices and energy conservation techniques. Their transmitting and receiving range is more as compared to other sensors up to 100 meters. They do not require coming in near contact to identify the device rather they identify devices from a distance [114, 118, 131, 138–143]. Frequency range of the RFID is divided into four categories:

– Low Frequency (135 KHz or less)
– High Frequency (13.56 MHz)
– Ultra-High Frequency (862–928 MHz)
– Microwave Frequency (2.4–5.80 GHz).

- **Passive Tag**: These tags get power supply from the energy harvesting technique. When an electric field or the beam is radiated from the RFID reader from that electric field power is generated [114, 118, 131, 138–143]. The power requirement is very low for these passive tags it will consume only 30 μW. Passive Tags are low cost and their production is more as compared to others. These tags also consist of EPC which can be readable with the help of RFID readers.

(iii) **Wireless Sensor Network**: wireless sensor network plays a very crucial role in the field of information communication. The basic structure of the wireless sensor network is, it contain wireless sensors with good configuration as they have an on-board processor, battery, sensor, communication system that is deployed randomly in a typical region or in worst conditions where human reach is not possible and not good for health. Hundred of sensors make a network and start collecting information regarding regions of interest, after collecting information they send that data to the base station which is located at another place with continuous power supply [118, 137, 144, 145]. Some sensors are temperature sensors, humidity sensors, vibration sensors, spy sensors etc. They are working over IEEE 802.15.4 standards.

(iv) **Bluetooth**: Bluetooth is a low priced technology that removes the wires between devices. Because of its short range, effective and low cost it is used very frequently for short communication. Devices like cameras, mobile phones, PC's peripheral etc use Bluetooth in short transmission range and with uninterrupted connection. Normally the range of Bluetooth device is 10 to 50 meters. It is first implemented by Ericsson Mobile Company in the year 1994. It creates a personal area network (PAN) and it works on IEEE 802.15.1 standards [118, 135, 146].

(v) **ZigBee**: It is a protocol that was designed to strengthen the WSN. It was first proposed in year 2001 by ZigBee Alliances. The ZigBee protocol is a Low power wireless network protocol that works on IEEE 802.15.4 standards. This protocol is low cost, low communication range, reliable and flexible in nature.Thetransmission range is about 100 meters with 250 kbps bandwidth. Main applications where ZigBee protocol is implemented are in agriculture sector, home automation, health care sector, smart buildings [118, 137, 139, 147].

(vi) **Near Field Communication(NFC)**: NFC is a "short-range wireless" technology which operates at the frequency of 13.56 MHz. It demands only 3–5 cm distance. This technology improves the lifestyle of humans by making transition simple and easier, exchanging digital content with speed and makes connection between physical devices easier with one tap; line of sight does not required [146, 148–150].

(vii) **Actuators**: These are the devices that are used to convert energy into motion.The main example is, hydraulic lift, ac to dc converter, use of water stream to generate electricity, use of gas pressure to generate electricity, stepper motor [109, 146, 151–153].

(viii) **Wireless Fidelity (Wi-Fi)**: Wi-Fi is a network topology that permits devices to connect and communicate information over a wireless link. Father of Wi-Fi is Vic Hayes. He developed this technology in 1991 at Netherland. Wi-Fi Connection has a speed up to 100 Mbps. It creates "Wireless Local Area Network (WLAN)" that connects offices, homes, traffic, airports, bus-stands, factories etc. It works on IEEE 802.11 standards [135–137, 139, 141–143, 152, 154].

2.3 Applications of IOT

Internet of things is now an important part of our life. Every day we are familiar with many smart device moreover, we are using many smart devices in our daily routine. Internet has made everything easy and information is available on our figure tips [108, 155–159]. It was predicted in late 1990's that by 2020, billions of users would be connected to the internet and this assumptions came true. Internet of things offers a number of applications to make life much easier and hassle free like smart home, smart city, smart transport system and many more [155, 157, 160–168]. Some of the applications are mentioned below:

(i) **Home Automation**: Intelligent and automated services are provided by the sensors that are deployed in the house. If a person leaving his house and he forgets to switch off the lights or fans and any other electronic appliance then these gadgets are much smarter and if they find no movement in their sensing range, they will automatically get turned off. In this way they can save energy consumption. Energy conservation is done smartly in smart houses by the help of smart sensors [169–175]. Let us suppose it's very hot outside, and AC is running continuously at a lower temperature. By making AC smart enough they will sense the room temperature and once it achieves the room temperature they will get turned off. Moreover, IOT plays an important role in the design of the house, there are vibration sensors are available in market, by sensing the vibration they will tell about the condition of the architecture weather it needs repair or not.

(ii) **Smart Cities**: Smart cities are one of the visions of IOT, in a city there are requirements of well planned structure of roads, parks, living area, sewage system, transport system etc. To implement this vision smartly using Internet of Things, everything should be smart and connected to the internet [129, 170, 176–187]. Smart transport systems can manage routine traffic using sensors. If there is any work in progress on any road, sensors will communicate to everyone to divert their routes and choose a best way to reach the destination. If any accident takes place on the road, sensors are capable of sending an immediate report to the ambulance so that life can be saved by giving first aid treatment. The lighting systemson the side of roads are much smarter so that, they can provide lighting moreover they can save the energy also. Intelligent parking systemsare an important component of smart cities. Everybody can check with the help of internet for available parking.

(iii) **Health and Fitness**: Health care applications demonstrate their benefits in improving health and make possible of independent living for elders and patients. These days many wearable devices embedded with internet are available in market. These devices take a full track of the health of patient and old age people [165, 180, 188–192]. They grab all the information's and data related to the health for example blood pressure, heart beats, sugar level, ECG reports and forward all recorded data to the respected doctor. As a result, doctorsmaintain constant vigilance over their patients and if any mixhap occurs, the doctor can immediately monitor and cure the patient life.

(iv) **Supply Chain**: Today IOT tries to remove the complexities from the real world problems that we face in our daily life. IOT is also used very frequently in business world. IOT uses RFID tags and NFC's to track the goods in supply chain. As a bulk of goods are present in the supply chain so to track and retrieve real time information these sensors are used. RFID tags are attached to goods makeingit easier to identify the goods because each good carries a unique id the supply chain [106, 193–200]. If the tags are embedded with geographical locations, it will improve the accuracy of findinga particular object in world.

(v) **Smart Agriculture**:Agriculture is currently experiencing a technological boom as a result of technological advancements.. Farmers are very aware of the technologies they are implementing the ongoing trends and technologies in their fields to make their work efficient and to produce more and more crops [201–210]. These all can be implemented with the help of sensors. Farmers are using green house experiments to improve their production by measuring the required temperature for the crop. Sensors are deployed inside the green house, they will sense the outside temperature and according to that, they will adjust the inside temperature. Many modified machines are also available to make farming easy. They will reduce the human cost and make profit to the farmers [211].

3 Related Work: Integration and Implementation of WSN with IOT Technology Against COVID-19

Al-Shalabi et al. [125], proposed a mechanism for monitoring symptoms of Covid-19 isolated patients for the age group of 60 years and above who came in contact with conformed cases with the help of Internet of things and wireless sensor networks. As World Health Organisation (WHO) declared this a pandemic because it spreads rapidly among communities by means of sneezing, coughing, contact with infected people. WHO addressed some common features of the Covid-19 as fever, dry cough, and sore throat. According to the author, person with the medical conditions or above the age 60 years might have more risk as compared to younger people. Therefore it is necessary to monitor those people and ask them to be self isolation for a period of 14 days. In this case, the author considers themost common symptom to be arise in temperature whichis measured with the help sensor. The proposed mechanism has a LM35 temperature sensor which is connected with the Arduino Uno Board. Sensor senses the temperature periodically and produces an analog output which is further converted into corresponding digital value with the help of Arduino Uno Board. After that, ESP8266 Wi-Fi module provides internet connectivity to the microprocessor with the use of Wi-Fi module readings is shared to a virtual cloud that acts like a server. On the other side, Doctor can access reading from any corner of the world and can provide medical assistance to the patient.

Maghded et al. [126], designed an AI-enabled framework for Covid-19 diagnoses using inbuilt smartphone sensors. WHO already announced coronavirus as a pandemic disease which is first located inWuhan city inChina. Virus can cause lung inflammation, fever, cough and pneumonia. These days lots of research is going on and researchers are striving to restrict the spread of Covid-19. Moreover, various machines and smart tools are being developed for the detection of coronavirus and to provide medical analysis. Covid-19 patients exhibit symptoms such as fever, fatigue, breathing difficulties. Some medical healthcare kits are present in market but they have a high cost and are difficult to install like CT scan, NAT, Blood samples. To overcome this issue, Maghdid et al designed a mechanism to detect Covid-19 using smart phone integrated sensors using Artificial Intelligent. In this technological era everybody is using smartphones hence there is no need to purchase any other clinical kit. It offers a low cost, easy to use and many problems to one solution. Author makes use of integrated sensor like, temperature is measured by fingerprint sensor, fatigue level is accessed by camera sensor, and microphone sensor has been utilized for cough voice sample. Algorithms are independently implemented on all of these sensors. By implementing machine learning techniques on input data, predictions are calculated.

Dong et al. [130], designed a Fog-Cloud-IOT platform [124, 212] and made a survey on prevention and control of COVID-19 using IOT platform. In the designed platform four layers are introduced that contains Perception Layer, Network Layer, Fog Layer, and Cloud Layer. Internet of Things has a wide range and uninterrupted connectivity due to this it can perform specific tasks and can achieve desired results.

The main motivation of this survey is to discuss the potential of IOT that can be implemented through various non pharmaceutical interventions (NPI's). The author has discussed some non pharmaceutical interventions like Breath Monitoring, Blood Oxygen Saturation Monitoring, Body Temperature Monitoring, Quarantine Monitoring, Contact Tracing and Mutation predictions.

Chowdhury et al. [127], proposed a technique for the diagnosis of COVID-19 patients with the help of digital chest X-ray images which is implemented by pre-trained deep CNN algorithms [213]. Moreover, author has described "Reverse Transcription Polymerase Chain Reaction (RT-PCR)" test which is very costly and a lazy process. It takes minimum of 2 days to generate report of the samples that cannot bearable at this peak time. In addition to this, it requires medical professional to operate these kits. It also requires a separate lab to store the samples of suspected people. In this paper, author has taken a big data base which is a combination of 423 COVID-19, 1485 viral pneumonia and 1579 normal chest X-ray images. By implementing image augmentation, transfer learning techniques are used to instruct deep CNN [214]. For two classification schemes, there are 8 deep learning models were trained. One model is utilized to analyse X-ray and COVID -19 images whereas the other model is utilized to analyse normal, viral and Covid-19 pneumonia images [215, 216]. Both the classification models are executed with and without image augmentation techniques. At last, the experimental results are very interesting and with high accuracy.

Ndiaye et al. [128], presents a survey on how the pandemic COVID-19 affects the IOT technology? As the World Health Organisation (WHO) already declared novel coronavirus as a pandemic, virus hasan adverse effect onthe world economy and people. Many people have already lost their jobs and lives due to this pandemic. Researchers are continuously developing new techniques to identify the affected people and to cure the spread of this virus. Moreover many of the technologies are being invited with the help of artificial intelligence and deep learning [122, 217],which make it easy to track and monitor the affected people but researchstill is going on and it's become need of hour. The author's main motivation in this survey is to usenewly designed techniques by utilizingInternet of Things and machine learning techniques. The author also describes the ecosystem for the COVID-19. In addition to this, author also tells the importance of Health- Internet of Things (H-IOT) [218].

Ali et al. [137], focuses on the role of smart Wireless Medical Sensors Network (WMSN) during the period of COVID-19 pandemic. According to the author, the adaptation of sensors in our daily life is an important tool in the present scenario. As the person whohas mild or swearssymptoms of COVID-19, in their quarantine period wireless medical sensors plays an important role. As it reduces human interaction and helps doctor to monitor it from their clinic or home. In addition to this, author has discussed the common symptoms of the coronavirus like temperature, respiratory problem, oxygen saturation and heartbeat. The main research objective of the authors study is to reduce the physical interaction between patient and doctor, nurse or other medical staff members because these are communicable disease and the main cause behind the spread of communicable disease is physical interaction as in the case of COVID-19, doctor suggests to make social distancing. Also author proposed a smart device that will overcome this issue, author make use of microcontroller with sensors to retrieve the information about the patients health 24 * 7 at any place. The proposed device is able to measure the temperature, heartbeats, respiratory problem, and oxygen saturation with the help of sensors. After that, collected data is processed by the microcontroller and is shared with the doctor.

Costanzo et al. [136], designed a "Non-Contact Integrated Body-Ambient Temperature sensor for COVID-19 pandemic". The infrared thermometer and humidity sensor are integrated to detect the human body temperature contactless. As COVID-19 is a communicable disease, it can spread while coming in contact with affected person. To avoid this contact, it requires wireless sensors with IOT platform to make it more reliable. In addition to this, author makes use of an Infrared sensor for the measurement of human body temperature. It contains a lens that will filter the infrared radiation and focus it on photoresist temperature sensor. Followingthat, the temperature sensor generates ananalog output thatis fed into the arduino board, whichconverts the analog input into corresponding output. On the other hand, a Relative Humidity (RH) sensor is used to provide voltage at the output and this voltage is also fed into arduino board. After molding the voltage RH and ambient temperature with body temperature is displayed onthe LED. Author also gives calculations for the Relative Humidity calculation, ambient temperature calculation and Body Temperature calculations.

Silva et al. [135], reviews the applications of wearable bioelectronics for patient monitoring and domiciliary hospitalization. In addition to this, author discussed about the internet of medical things (IOMT) that are more popular in this era.There are challenges and benefits ofthe hassle free monitoring of patients from their home. "Digital healthcare" and "Domiciliary Hospitalization" are fundamental key elements of future healthcare. This technique can improve the quality of life and can helpindividuals fight chronic diseases. Through this doctor or any medical person can monitor the health status of patients 24 * 7. "Domiciliary Hospitalization" means a person is hospitalized at their home with all time medical care monitoring. Nowadays, medical facilities are not on the marks because of rapid increase in number of patients affected byCOVID-19. There is a high requirement of medical beds, equipments and person to person care. To overcome this issue, many countries already implement the home isolation and some countries are now applying home quarantine to get rid of this coronavirus. To make this possible, wireless sensor networks and Internet of medical things are integrated with each other toprovide a good healthcare system. Author also discussed the relationship of these technologies with the ongoing pandemic. The main focus of the study is to highlight the wearable Bioelectronics equipments with the integration of Internet of Medical things [219]. Nowadays many body-interfacing devices are available in the market like wearable watches, fitness trackers, these devices are used to acquire and transmit the "electrophysiological data". Some of the electrophysiological sensing devices are discussed by the author, like "skin interface electrophysiological sensing" as our body is radiating various signals that consists of various significant information regarding out body. "Biopotential" signals consist of electric signals that are generated by our heart, brain and muscles. Hence, to measure these electric signals, Electrocardiogram (ECG), Electroencephalogram (EEG) and Electromyography (EMG) are used. Bio impedance sensors are used to measure impedance of a particular part of the body. Bio acoustics are implemented to measure the sound produced by the body organs. Moreover, the author providesa description of theIOMT system's implementation, which includes three different layers, the sensors layer, which is equipped with wireless body sensors, the communication layer, which consists of devices, access points, and communication technologies such as 3G/4G WiFi, and the application layer, which is responsible for IOMT system operations.

Islam et al. [142], designed a "smart healthcare monitoring system using IOT". The Internet of Things plays an important role in the field of health care monitoring. we can make our devices smart and controllable from a distanceby effectively utilizing IOT technology. Today's scenario is worst in conditions and every part of the world is suffering from the COVID-19 pandemic. Millions of people havelost their life and many are suffering from the virus. It hasan adverse effect on the economy and the medical facilities. Author has designed a device that is used for the healthcare monitoring. Proposed device is an energy efficient and flexible interface. Author considered some factors for the monitoring, body temperature, pulse rate, some dangerous gases like CO and CO_2. The WiFi module is also used to track these factors and transfer information about patients' health. ESP32 microcontroller is used to make computation over the gathered data from the sensors. Heart beat sensor is utilized to track the pulse rate of the heart that lies between 60 and 100 beats per minutes. LM35 body temperature sensor is utilized for the optimized temperature. It takes reading in Kelvin's. DHT11 room temperature sensor is utilized for the calculation of the room temperature. MQ-9 CO sensor is utilized for the detection of CO and CH_4 gases. MQ-135 CO_2 sensor is used for air quality control system. The procedure of the health care monitoring system depends upon three layers, sensor module, data processing module and web user interface. All the sensors collect their information and forward it to the microprocessor for further computations. With the help of WiFi module the stored information is forwarded to the ThingSpeak virtual cloud [220, 221]. It makes easy to access the information of the patient and can see the values in the form of graphs.

Anifah et al. [141], proposed an "Internet of Things-based Monitoring System of Patients using W1209 Digital Thermostat and Pulse Sensor". This proposed system is an example of e-health technologies. The author has utilized the system for peoplewho areat a period of home or self quarantine. Many of the smart devices are already present in the market but still the advancements are going on in the field of health care. Healthcare is a burning matter nowadays, due to the pandemic COVID-19. The author takes two parameters to examine, body temperature and heart beat rate. As the increase in body temperature is a very common symptom of the coronavirus, so it is easy to make a guess that the person has possibility to being affected with COVID-19. The methodology is given as, W1209 Digital Thermostat for temperature sensing and for the support of Internet of Things WEMOS DI mini is implemented with a WiFi module. Both the sensors are connected to the WEMOS DI mini and with the use of WiFi data is transferred to the cloud and it can be retrieved anytime by medical staffs to examine the condition of the patient who isin their quarantine period.

Ghorbel et al. [143], design an Intelligent Medical Bracelet (IMB) with the help of IOT technology [222]. Author has design this device bracelet for the cure of COVID-19 pandemic. Wireless sensor network make improvements in the last decades as they are integrated with the IOT that boosts the utility of the sensors. Moreover, various technologies are present nowadays to diagnose diseases but they are costly and they require professional skills to operate the equipment. It is the need of hour to have small and user friendly equipments, so that person can operate those devices. In addition to this, author designed smart equipment named as IMB for the measurement of body temperature, heart beat of the affected person. The targeted population are children, elders and disabled. They all can easily access the device due to its user friendly interface and easy to use. This bracelet has many functionality such as display measured temperature, display number of heart beats, it offers 3 types of alert system: sound, light and an SMS service when the health is in anunstable state.

4 Comparison Between Various Already Designed Architectures for the Prevention of COVID-19

See (Table 1).

5 Conclusion

The COVID-19 is a chronic disease which is spreading all over the world. Many people have lost their lives and many are fighting for their lives in hospitals. It affects nature and harms humanity. World Health Organisation has declared COVID-19 as a pandemic. Many researches are going in the state of art pandemic but till date no pharmaceutical vaccine is available in the market. To cure the spread of corona virus many non-pharmaceutical interventions are invented with the help of technologies that already exist like artificial intelligence, machine learning, wireless sensors and internet of things. Many advertisements are being done to make people aware of precautions to stop the spread of viruses like social distancing, hand washing, wearing masks and using sanitizer. This survey concludes the ongoing advancements in the technology using wireless sensor networks and internet of things for the cure of COVID-19.

In the present scenario, whole world is fighting against COVID-19 pandemic. E-health care is a sector of opportunities there are many more things to do and can be implemented with the collaboration of artificial intelligence and machine learning techniques. With the integration of wireless sensor networks and internet of things, many more effective products and techniques can be designed/proposed that will be reliable, easy to operate, cost effective and with the high percentage of accuracy. In addition to this, risk of privacy and security should be taken into consideration due

Table 1 Comparison between various already designed architectures for the prevention of COVID-19

Author	Paper title	Objective/focus	Contribution	Technique used	Limitation
Al-Shalabi et al. [125]	COVID-19 Symptoms Monitoring Mechanism using Internet of Things and Wireless Sensor Networks	Designed mechanism is implemented for the people who are in quarantine period	Design & Simulation of the COVID-19 symptoms monitoring mechanism. Remote monitoring of patient health data is being done. Main focus is on Elder people	Temperature is measured with the help of LM35 Temperature sensor with integration of Arduino Uno Board. ThingSpeak IoT server is used for data storage	LM35 temperature sensor's readings are not accurate. Real time implementation is not shown
Maghded et al. [126]	A Novel AI-enabled Framework to Diagnose Coronavirus COVID-19 using Smartphone Embedded Sensors: Design Study	Artificial Intelligence equipped framework is implement-ed with the help of smart phone sensors to calculate the severity of disease	New framework is proposed for detection of COVID-19 by using in-built smartphone sensors. It is a cost effective solution. Common people can also use this framework	Figure print sensor takes input for the temperature check. Microphone sensor takes cough sample. CT scan images of lungs are captured by camera sensor. After taken input from different sensors comparison are made with already stored data. By applying CNN or RNN machine learning technique predictions are made weather person is affected or not	COVID-19 disease can effect anyone on the planet, data stored for the comparison is of old aged person (61 years old), people with different age group has different body structures hence, comparison data should have of different age groups so that diagnose can be made easily. Algorithm is not mentioned

(continued)

Table 1 (continued)

Author	Paper title	Objective/focus	Contribution	Technique used	Limitation
Anifah et al. [141]	Internet of Things based monitoring system of patients using W1209 Digital thermostat and Pulse sensor	Proposed an e-health technology that will help medical staff for patient assistance	Technology contributes in accessing body temperature and beating rate of heart	Testing of temperature is done with the help of W1209 Thermostat and heart beat rate is countered by Beat Per Minute (BPM) Sensor	Designed algorithm is not well explained and flow chart is missing. Series of BPM sensor is not mentioned
Islam et al. [142]	Development of smart healthcare monitoring system in IOT Environment	Main focus is on the patient's health monitoring in hospital	A smart health care system is designed for the observation of patient current health in hospital and room monitoring	Tracking heart beat, body temperature, room temperature. Various sensors are integrated with ESP32 module. ThingSpeak online web server is utilized for the real time observation	Cost of implementation is not mentioned. Architecture is explained but algorithm and flow chart is missed
Divya et al. [223]	Smart health care monitoring based on Internet of things (IOT)	The main focus is on tracking vital parameters of isolated patients	Remote Patient Health Monitoring System (RPHMS) is proposed. Body temperature, blood pressure, heart beat and ECG are detected	Arduino NANO, BMP-180 pressure sensor for BP measurement, DHT11 sensor for temperature and humidity sensing, AD8232 ECG module is implemented	Methodology is explained but it has a complex structure. Production cost is not defined

(continued)

Table 1 (continued)

Author	Paper title	Objective/focus	Contribution	Technique used	Limitation
Bhuyan et al. [224]	Design and simulation of heartbeat measure-ment system using arduino micro-controller in proteus	The main focus of the work is to plot a model and simulate heart beat measurement	There are five different case studies of the HRM model i.e. new born baby, older children, teenagers, adult and active athlete	Heart beat sensor is integrated with the Arduino UNO R3. Readings are taken on LCD screen. Whole project is implemented on a PCB design. Proteus software is utilized for the better placement of the components	Comparison is not done with existing modules. Author claims that module is cost effective but it is not mentioned about the expenses
Rahimoon et al. [225]	Design of a contactless body temperature measure-ment system using arduino	To develop a Human Body Temperature tracking system to cure the COVID-19 spread	Body temperature has been accessed with minimum error. 15'C difference is between both the temperature sensors	In present scenario two different temperature sensors are utilized i.e. LM35 and MLX-90614 sensors. With the help of ESP-WiFi module readings are real time monitoring is done	Work can be extended by adding more accessing features. Web server name is not mentioned. Furthermore there is no real time graphs attached in results
Abuzairi et al. [226]	Infrared thermometer on the wall (iThermo-wall): An open source and 3-D print infrared thermometer for fever screening	The design and implement IR thermometer that will be mount on a wall to reduce the impact of Covid-19 infected people to come in contact	A well designed hardware product is introduced named as ithermo-wall. Well defined process of formation and production cost is all mentioned i.e. $35 USD	MLX 90614 temperature sensors is used with GY-906 module, Arduino Nano micro-controller is implemented, two lithium ion battery, OLED display for displaying results, buzzer for the indication are used	Product is good but it is not cost effective as $35 is cost. Working algorithm and flow chart is missing

(continued)

Table 1 (continued)

Author	Paper title	Objective/focus	Contribution	Technique used	Limitation
Hasan et al. [227]	Designing ECG monitoring healthcare system based on Internet of things Blynk application	Proposed system is designed for ECG tracking	Designed product is easy to use. Anybody can have access to their ECG and can communicate with doctor without attending physical counselling	ECG probes are connected to AD8382 ECG sensor. Arduino Uno is base for implementation. With the help of Blynk application ECG readings are shown on smart phones	It totally depends upon the user how they will place the ECG probes. Readings can fluctuate. Security is major concern
Hadis et al. [228]	IoT based patient monitoring system using sensor to detect, analyse and monitor two primary vital signs	To identify, examine and monitor crucial signs body temperature and respiratory rate	Patient monitoring system is designed for the patients admitted in hospitals or at their quarantine period at home. Proposed system is useful for the nurses & doctors	Two LM35 and one MLX90614 temperature sensors are implemented. Arduino Mega 2560 micro-controller is used. With the help of Blynk application readings are taken out on nurse smart phones	Sensor LM35 is not reliable for accurate readings that may lead to unhappening
Sam et al. [229]	Progressed IOT based remote health monitoring system	The main focus is to design remote health monitoring system	Tracking of pulse rate, circulatory strain, breath rate, body temperature, saline dimensions	Implementation of the proposed work is done with the help of Arduino UNO. Different sensors are used for different types of measurement. Output is taken on iot web server	Sensor modules are not mentioned. Algorithm is not defined

(continued)

Table 1 (continued)

Author	Paper title	Objective/focus	Contribution	Technique used	Limitation
Mohammad et al. [230]	Novel COVID-19 detection and diagnosis system using IOT based smart helmet	The main objective of the model is to track the COVID-19 effective people by implementing thermal imaging	A system is designed for screening using smart helmet with thermal image camera mounted over it	Smart helmet consist of two cameras; one is for thermal scanning and other is optical camera. Face detection is done with the help of cascade classification. Google location history is used for the identification of the visiting places	No implementation cost is mentioned. Major security concerns
Mhatre et al. [231]	Non invasive E-Health care monitoring system using IOT	The main objective of the research is to design a E-health care system that can be operated remotely with the help of IOT	The proposed model is for the Non-invasive patients	Pulse sensor, temperature sensor, blood pressure sensor and glucose sensor are integrated with Arduino UNO micro-controller. Thing Speak web server is utilized for the remote observation of the data	Data security is major concern. LM35 sensor is used that gives readings with >5 difference
Petrovic et al. [232]	IoT based system for COVID-19 Indoor safety monitoring	Provides affordable IoT based solution for improving indoor safity from COVID-19	Contactless temperature sensing, Mask detection and Social distancing checks are the main aspects of the proposed model	Message Queuing Telemetry Transport (MQTT) are used for M2M communication MLX90614 temperature sensor is used. Raspberry Pi microcontroller is utilized	Haarcascades mask/ face/ full body detection algorithm are implemented. Distancing check accuracy is 65–73%. Top notch Laptop configurations are required for properly implementation of the model. Not cost effective

to increase in traffic over internet. All devices are connected to web and transferring of personal data takes place that can lead to major threat for mankind.

References

1. Raghavendra, C.S.: K M Sivalingam. Wireless sensor networks, Z.T. (2006)
2. Akyildiz, I.F., Su, W., Sankarasubramaniam, Y., Cayirci, E.: Wireless sensor networks: a survey. Comput. Netw. **38**(4), 393–422 (2002). https://doi.org/10.1016/s1389-1286(01)00302-4
3. Tree, S.: Wireless sensor networks. Self **1**(R2) (2014)
4. Akyildiz, I.F., Vuran, M.C.: Wireless Sensor Networks, vol. 4 (2010)
5. Mainwaring, A., Culler, D., Polastre, J., Szewczyk, R., Anderson, J.: Wireless sensor networks for habitat monitoring. In: Proceedings of the 1st ACM International Workshop on Wireless Sensor Networks and Applications, pp. 88–97 (2002)
6. Estrin, D., Girod, L., Pottie, G., Srivastava, M.: Instrumenting the world with wireless sensor networks. In: 2001 IEEE International Conference on Acoustics, Speech, and Signal Processing. Proceedings (Cat. No. 01CH37221), vol. 4, pp. 2033–2036 (2001)
7. Guy, C.: Wireless sensor networks. In: Sixth International Symposium on Instrumentation and Control Technology: signal Analysis, Measurement Theory, Photo-electronic Technology, and Artificial Intelligence, vol. 6357 (2006)
8. Pottie, G.J.: Wireless sensor networks. In: Information Theory Workshop (Cat. No. 98EX131) pp. 139–140 (1998)
9. Stankovic, J.A.: Wireless sensor networks. Computer **41**(10), 92–95 (2008). https://doi.org/10.1109/mc.2008.441
10. Fong, D.Y.: Wireless Sensor Networks. Internet of Things and Data Analytics Handbook, pp. 197–213 (2017)
11. Dargie, W., Poellabauer, C.: Fundamentals of Wireless Sensor Networks: theory and Practice (2010)
12. Townsend, C., S.A.: Wireless Sensor Networks, vol. 20, pp. 15–21 (2005)
13. Zhao, F., Guibas, L.J., Guibas, L.: Wireless Sensor Networks: an Information Processing Approach (2004)
14. Arampatzis, T., Lygeros, J., Manesis, S.: A survey of applications of wireless sensors and wireless sensor networks. In: Proceedings of the 2005 IEEE International Symposium on, Mediterrean Conference on Control and Automation Intelligent Control, pp. 719–724 (2005)
15. Yang, K.: Wireless Sensor Networks (2014)
16. García-Hernández, C.F., Ibarguengoytia-Gonzalez, P.H., J.G.H.J.A.P.D.: Wireless sensor networks and applications: a survey. IJCSNS Int. J. Comput. Sci. Netw. Secur. **7**(3), 264–273 (2007)
17. Hailing, C.L., Yong, M., Tianpu, L., Wei, L., Ze, Z.: Overview of wireless sensor networks. J. Comput. Res. Dev. **42**(1), 163–163 (2005)
18. Estrin, D., Sayeed, A., Srivastava, M.: Wireless sensor networks. In: Tutorial at the Eighth ACM International Conference on Mobile Computing and Networking, vol. 255 (2002)
19. Ko, J., Lu, C., Srivastava, M.B., Stankovic, J.A., Terzis, A., Welsh, M.: Wireless sensor networks for healthcare. Proc. IEEE **98**(11), 1947–1960 (2010). https://doi.org/10.1109/jproc.2010.2065210
20. Li, Y., T.M.T.: Wireless Sensor Networks and Applications (2008)
21. Karl, H., Willig, A.: Protocols and Architectures for Wireless Sensor Networks (2007)
22. Sohraby, K., Minoli, D., Znati, T.: Wireless Sensor Networks: technology, Protocols, and Applications (2007)
23. Roundy, S., Steingart, D., Frechette, L., Wright, P., Rabaey, J.: Power sources for wireless sensor networks. In: European Workshop on Wireless Sensor Networks, pp. 1–17 (2004)

24. Benadda, M., Belalem, G.: Improving road safety for driver malaise and sleepiness behind the wheel using vehicular cloud computing and body area networks. Int. J. Softw. Sci. Comput. Intell. (IJSSCI) **12**(4), 19–41 (2020)

25. Buratti, C., Conti, A., Dardari, D., Verdone, R.: An overview on wireless sensor networks technology and evolution. Sensors **9**(9), 6869–6896 (2009). https://doi.org/10.3390/s90906869

26. Hande Alemdar, C.E.: Wireless sensor networks for healthcare: a survey. Comput. Netw. **54**(15), 2688–2710 (2010). https://doi.org/10.1016/j.comnet.2010.05.003

27. Romer, K., Mattern, F.: The design space of wireless sensor networks. IEEE Wirel. Commun. **11**(6), 54–61 (2004). https://doi.org/10.1109/mwc.2004.1368897

28. Corke, P., Wark, T., Jurdak, R., Hu, W., Valencia, P., Moore, D.: Environmental wireless sensor networks. Proc. IEEE **98**(11), 1903–1917 (2010). https://doi.org/10.1109/jproc.2010.2068530

29. Misra, S., Misra, S.C., Woungang, I.: Guide to Wireless Sensor Networks (2009)

30. Fahmy, H.M.A.: Wireless Sensor Networks (2020)

31. Porter, J., Arzberger, P., Braun, H.W., Bryant, P., Gage, S., Hansen, T., Hanson, P., Lin, C.C., Lin, F.P., Kratz, T., et al.: Wireless sensor networks for ecology. BioScience **55**(7), 561–561 (2005). https://doi.org/10.1641/0006-3568(2005)055[0561:wsnfe]2.0.co;2

32. Slijepcevic, S., Potkonjak, M.: Power efficient organization of wireless sensor networks. In: ICC 2001. IEEE International Conference on Communications. Conference Record (Cat. No. 01CH37240), vol. 2, 472–476 (2001)

33. Chen, Y., Zhao, Q.: On the lifetime of wireless sensor networks. IEEE Commun. Lett. **9**(11), 976–978 (2005)

34. Hailing, C.L.J., Yong, M., Tianpu, L., Wei, L., Ze, Z.: Overview of wireless sensor networks. J. Comput. Res. Dev. **1**, 21–21 (2005)

35. Zheng, J., Jamalipour, A.: Wireless Sensor Networks: a Networking Perspective (2009)

36. Kuorilehto, M., Hännikäinen, M., T.D.H.: A survey of application distribution in wireless sensor networks. EURASIP J. Wirel. Commun. Netw. **2005**(5), 1–15 (2005). https://doi.org/10.1155/wcn.2005.774

37. Boukerche, A., Oliveira, H.A., Nakamura, E.F., Loureiro, A.A.: Localization systems for wireless sensor networks. IEEE Wirel. Commun. **14**(6), 6–12 (2007)

38. Khemapech, I., Duncan, I., Miller, A.: A survey of wireless sensor networks technology. In: 6th Annual Postgraduate Symposium on the Convergence of Telecommunications, Networking and Broadcasting, vol. 13 (2005)

39. Isabel Dietrich, F.D.: On the lifetime of wireless sensor networks. ACM Trans. Sens. Netw. **5**(1), 1–39 (2009). https://doi.org/10.1145/1464420.1464425

40. Al-Karaki, J.N., Kamal, A.E.: Routing techniques in wireless sensor networks: a survey. IEEE Wirel. Commun. **11**(6), 6–28 (2004). https://doi.org/10.1109/mwc.2004.1368893

41. Wang, Q., Balasingham, I.: Wireless sensor networks-an introduction. In: Wireless Sensor Networks: application-Centric Design, pp. 1–14 (2010)

42. Jin, J., Gubbi, J., Marusic, S., Palaniswami, M.: Secure and optimized load balancing for multitier IoT and edge-cloud computing systems. IEEE Internet Things J. **8**(10), 8119–8132 (2021). https://doi.org/10.1109/jiot.2020.3042433

43. Elgendy, I.A., Muthanna, A., Hammoudeh, M., Shaiba, H., Unal, D., Khayyat, M.: Advanced deep learning for resource allocation and security aware data offloading in industrial mobile edge computing (2021)

44. Elgendy, I.A., Zhang, W.Z., Zeng, Y., He, H., Tian, Y.C., Yang, Y.: Efficient and secure multi-user multi-task computation offloading for mobile-edge computing in mobile IoT networks. IEEE Trans. Netw. Serv. Manag. **17**(4), 2410–2422 (2020)

45. Elgendy, I.A., Zhang, W., Tian, Y.C., Li, K.: Resource allocation and computation offloading with data security for mobile edge computing. Future Gener. Comput. Syst. **100**, 531–541 (2019). https://doi.org/10.1016/j.future.2019.05.037

46. Hill, J., Horton, M., Kling, R., Krishnamurthy, L.: The platforms enabling wireless sensor networks. Commun. ACM **47**(6), 41–46 (2004). https://doi.org/10.1145/990680.990705

47. Baronti, P., Pillai, P., Chook, V.W., Chessa, S., Gotta, A., Hu, Y.F.: Wireless sensor networks: a survey on the state of the art and the 802.15.4 and zigbee standards. Comput. Commun. **30**(7), 1655–1695 (2007). https://doi.org/10.1016/j.comcom.2006.12.020
48. Tolle, G., Culler, D.: Design of an application-cooperative management system for wireless sensor networks. In: Proceedings of the Second European Workshop on Wireless Sensor Networks, pp. 121–132 (2005)
49. Hill, J.L.: System architecture for wireless sensor networks. Doctoral dissertation (2003)
50. Li, J., Ma, H., Li, K., Cui, L., Sun, L., Zhao, Z., Wang, X.: Wireless sensor networks: 11th china wireless sensor network conference. Revis. Sel. Pap. **812** (2017)
51. Cardei, M., Wu, J.: Coverage in Wireless sensor Networks: handbook of Sensor Networks, vol. 21, 201–202 (2004)
52. Abd EL-Latif, A.A., Abd-El-Atty, B., Abou-Nassar, E.M., Venegas-Andraca, S.E.: Controlled alternate quantum walks based privacy preserving healthcare images in internet of things. Opt. Laser Technol. **124**, 105942–105942 (2020). https://doi.org/10.1016/j.optlastec.2019.105942
53. Rashid, B., Rehmani, M.H.: Applications of wireless sensor networks for urban areas: a survey. J. Netw. Comput. Appl. **60**, 192–219 (2016). https://doi.org/10.1016/j.jnca.2015.09.008
54. Oliveira, Oliveira, L.M., Rodrigues, J.J.: Wireless sensor networks: a survey on environmental monitoring. J. Commun. **6**(2), 143–151 (2011). https://doi.org/10.4304/jcm.6.2.143-151
55. Kulkarni, R.V., Forster, A., Venayagamoorthy, G.K.: Computational intelligence in wireless sensor networks: a survey. IEEE Commun. Surv. Tutor. **13**(1), 68–96 (2011). https://doi.org/10.1109/surv.2011.040310.00002
56. Roundy, S., Wright, P.K., Rabaey, J.M.: Energy Scavenging for Wireless Sensor Networks, pp. 45–47 (2003)
57. Sharma, S., Bansal, R.K., Bansal, S.: Issues and challenges in wireless sensor networks. In: 2013 International Conference on Machine Intelligence and Research Advancement, pp. 58–62 (2013)
58. Wang, Y.O.N.G., Wang, Z.B.: Wireless sensor networks. In: ch. Compression Techniques for Wireless Sensor Networks, pp. 207–231 (2004)
59. Schurgers, C., Srivastava, M.B.: Energy efficient routing in wireless sensor networks. In: MILCOM Proceedings Communications for Network-Centric Operations: creating the Information Force (Cat. No. 01CH37277), vol. 1, pp. 357–361 (2001)
60. Chen, D., Varshney, P.K.: QoS support in wireless sensor networks: a survey. In: International Conference on Wireless Networks, vol. 233, pp. 1–7 (2004)
61. Durišić, M.P., Tafa, Z., G.D.V.M.: A survey of military applications of wireless sensor networks. In: Mediterranean Conference on Embedded Computing, pp. 196–199 (2012)
62. Alsheikh, M.A., Lin, S., Niyato, D., Tan, H.P.: Machine learning in wireless sensor networks: algorithms, strategies, and applications. IEEE Commun. Surv. Tutor. **16**(4), 1996–2018 (2014). https://doi.org/10.1109/comst.2014.2320099
63. Gad, R., Talha, M., Abd El-Latif, A.A., Zorkany, M., Ayman, E.S., Nawal, E.F., Muhammad, G.: Iris recognition using multi-algorithmic approaches for cognitive internet of things (ciot) framework. Future Gener. Comput. Syst. **89**, 178–191 (2018). https://doi.org/10.1016/j.future.2018.06.020
64. Ramson, S.J., Moni, D.J.: Applications of wireless sensor networks-a survey. In: 2017 International Conference on Innovations in Electrical, Electronics, Instrumentation and Media Technology (ICEEIMT), pp. 325–329 (2017)
65. Krishnamachari, L., Estrin, D., Wicker, S.: The impact of data aggregation in wireless sensor networks. In: Proceedings 22nd International Conference on Distributed Computing Systems Workshops, pp. 575–578 (2002)
66. Ojha, T., Misra, S., Raghuwanshi, N.S.: Wireless sensor networks for agriculture: the state-of-the-art in practice and future challenges. Comput. Electron. Agric. **118**, 66–84 (2015). https://doi.org/10.1016/j.compag.2015.08.011
67. Gungor, V.C., Lu, B., Hancke, G.P.: Opportunities and challenges of wireless sensor networks in smart grid. IEEE Trans. Ind. Electron. **57**(10), 3557–3564 (2010). https://doi.org/10.1109/tie.2009.2039455

68. Khan, S., Pathan, A.S.K., Alrajeh, N.A.: Wireless Sensor Networks: current Status and Future Trends (2016)
69. Swami, A., Zhao, Q., Hong, Y.W., Tong, L..: Wireless Sensor Networks: signal Processing and Communications Perspectives (2007)
70. Ganesan, D., Cerpa, A., Ye, W., Yu, Y., Zhao, J., Estrin, D.: Networking issues in wireless sensor networks. J. Parall. Distrib. Comput. **64**(7), 799–814 (2004). https://doi.org/10.1016/j.jpdc.2004.03.016
71. El Emary, I.M., Ramakrishnan, S.: Wireless Sensor Networks: from Theory to Applications (2013)
72. Stankovic, J.A., Wood, A.D., He, T.: Realistic applications for wireless sensor networks. Theor. Asp. Distrib. Comput. Sensor Netw. 835–863 (2011)
73. Anastasi, G., Conti, M., Di Francesco, M., Passarella, A.: Energy conservation in wireless sensor networks: a survey. Ad Hoc Netw. **7**(3), 537–568 (2009). https://doi.org/10.1016/j.adhoc.2008.06.003
74. Zhang, S., Zhang, H.: A review of wireless sensor networks and its applications. In: 2012 IEEE International Conference on Automation and Logistics, pp. 386–389 (2012)
75. Stankovic, J.A.: Research challenges for wireless sensor networks. ACM SIGBED Rev. **1**(2), 9–12 (2004). https://doi.org/10.1145/1121776.1121780
76. Nakamura, E.F., Loureiro, A.A., Frery, A.C.: Information fusion for wireless sensor networks. ACM Comput. Surv. **39**(3), 9–9 (2007). https://doi.org/10.1145/1267070.1267073
77. Barrenetxea, G., Ingelrest, F., Schaefer, G., Vetterli, M.: Wireless sensor networks for environmental monitoring: the sensorscope experience. In: IEEE International Zurich Seminar on Communications, pp. 98–101 (2008)
78. Tavares, J., Velez, F., Ferro, J.: Application of wireless sensor networks to automobiles. Meas. Sci. Rev. **8**(3), 65–70 (2008). https://doi.org/10.2478/v10048-008-0017-8
79. Winkler, M., Tuchs, K.D., Hughes, K., Barclay, G.: Theoretical and practical aspects of military wireless sensor networks. J. Telecommun. Inf. Technol. 37–45 (2008)
80. Min, R., Bhardwaj, M.: Low-power wireless sensor networks. In: Fourteenth International Conference on VLSI Design, pp. 205–210 (2001)
81. Wang, Y., Attebury, G., Ramamurthy, B.: A survey of security issues in wireless sensor networks (2006)
82. Hu, F., Cao, X.: Wireless Sensor Networks: principles and Practice (2010)
83. Mainetti, L., Patrono, L., Vilei, A.: Evolution of wireless sensor networks towards the internet of things: a survey. In: SoftCOM 2011, 19th International Conference on Software, Telecommunications and Computer Networks, pp. 1–6 (2011)
84. Lilia Paradis, Q.H.: A survey of fault management in wireless sensor networks. J. Netw. Syst. Manag. **15**(2), 171–190 (2007). https://doi.org/10.1007/s10922-007-9062-0
85. Mihaela Cardei, J.W.: Energy-efficient coverage problems in wireless ad-hoc sensor networks. Comput. Commun. **29**(4), 413–420 (2006). https://doi.org/10.1016/j.comcom.2004.12.025
86. Wheeler, A.: Commercial applications of wireless sensor networks using ZigBee. IEEE Commun. Mag. **45**(4), 70–77 (2007). https://doi.org/10.1109/mcom.2007.343615
87. Baggio, A.: Wireless sensor networks in precision agriculture. In: ACM Workshop on Real-World Wireless Sensor Networks (REALWSN 2005), vol. 20, pp. 1567–1576 (2005)
88. Shen, X., Wang, Z., Sun, Y.: Wireless sensor networks for industrial applications. In: Fifth World Congress on Intelligent Control and Automation, vol. 4, pp. 3636–3640 (2004)
89. Yu, Y., Rittle, L.J., Bhandari, V., LeBrun, J.B.: Supporting concurrent applications in wireless sensor networks. In: Proceedings of the 4th International Conference on Embedded Networked Sensor Systems, pp. 139–152 (2006)
90. Mahmood, M.A., Seah, W.K., Welch, I.: Reliability in wireless sensor networks: a survey and challenges ahead. Comput. Netw. **79**, 166–187 (2015). https://doi.org/10.1016/j.comnet.2014.12.016
91. Megerian, S., Koushanfar, F., Qu, G., Veltri, G., Potkonjak, M.: Exposure in wireless sensor networks: theory and practical solutions. Wirel. Netw. **8**(5), 443–454 (2002)

92. Dâmaso, A., Rosa, N., Maciel, P.: Reliability of wireless sensor networks. Sensors **14**(9), 15760–15785 (2014). https://doi.org/10.3390/s140915760

93. Manshahia, M.S.: Wireless sensor networks: a survey. Int. J. Sci. Eng. Res. **7**(4), 710–716 (2016)

94. Bokareva, T., Hu, W., Kanhere, S., Ristic, B., Gordon, N., Bessell, T., Rutten, M., Jha, S.: Wireless sensor networks for battlefield surveillance. In: Proceedings of the Land Warfare Conference, pp. 1–8 (2006)

95. Rajaravivarma, V., Yang, Y., Yang, T.: An overview of wireless sensor network and applications. In: Proceedings of the 35th Southeastern Symposium on System Theory, pp. 432–436 (2003)

96. Hadjidj, A., Souil, M., Bouabdallah, A., Challal, Y., Owen, H.: Wireless sensor networks for rehabilitation applications: challenges and opportunities. J. Netw. Comput. Appl. **36**(1), 1–15 (2013). https://doi.org/10.1016/j.jnca.2012.10.002

97. Han, G., Jiang, J., Shu, L., Niu, J., Chao, H.C.: Management and applications of trust in wireless sensor networks: a survey. J. Comput. Syst. Sci. **80**(3), 602–617 (2014). https://doi.org/10.1016/j.jcss.2013.06.014

98. Carlos-Mancilla, M., López-Mellado, E., Siller, M.: Wireless sensor networks formation: approaches and techniques. J. Sens. **2016**, 1–18 (2016). https://doi.org/10.1155/2016/2081902

99. Chand, S., Singh, S., Kumar, B.: Heterogeneous heed protocol for wireless sensor networks. Wirel. Pers. Commun. **77**(3), 2117–2139 (2014)

100. Raghunathan, V., Ganeriwal, S., Srivastava, M.: Emerging techniques for long lived wireless sensor networks. IEEE Commun. Mag. **44**(4), 108–114 (2006). https://doi.org/10.1109/mcom.2006.1632657

101. Heinzelman, W.R., Chandrakasan, A., Balakrishnan, H.: Energy-efficient communication protocol for wireless microsensor networks. In: Proceedings of the 33rd Annual Hawaii International Conference on System Sciences, p. 10 (2000)

102. Salim, A., Osamy, W., Khedr, A.M.: IBLEACH: intra-balanced LEACH protocol for wireless sensor networks. Wirel. Netw. **20**(6), 1515–1525 (2014). https://doi.org/10.1007/s11276-014-0691-4

103. Gupta, S., Marriwala, N.: Improved distance energy based leach protocol for cluster head election in wireless sensor networks. In: 4th International Conference on Signal Processing, Computing and Control (ISPCC), pp. 91–96 (2017)

104. Satapathy, S.C., Raju, K.S., J.K.M.V.B.: Proceedings of the Second International Conference on Computer and Communication Technologies: IC3T 2015, vol. 1 (2016)

105. Le, T.N., Pegatoquet, A., Berder, O., Sentieys, O.: Energy-efficient power manager and MAC protocol for multi-hop wireless sensor networks powered by periodic energy harvesting sources. IEEE Sens. J. **15**(12), 7208–7220 (2015). https://doi.org/10.1109/jsen.2015.2472566

106. Sethi, P., Sarangi, S.R.: Internet of things: architectures, protocols, and applications. J. Electr. Comput. Eng. **2017**, 1–25 (2017). https://doi.org/10.1155/2017/9324035

107. Dictionary: Dictionary. https://www.dictionary.com/browse/things. Accessed 09 May 2021

108. Madakam, S., Lake, V., Lake, V., Lake, V.: Internet of things (IoT): a literature review. J. Comput. Commun. **3**(05), 164–164 (2015)

109. Mattern, F., Floerkemeier, C.: From the internet of computers to the internet of things. In: From Active Data Management to Event-Based Systems and More, pp. 242–259 (2010)

110. Farooq, M.U., Waseem, M., Mazhar, S., Khairi, A., Kamal, T.: A review on internet of things (IoT). Int. J. Comput. Appl. **113**(1), 1–7 (2015). https://doi.org/10.5120/19787-1571

111. Wang, H., Li, Z., Li, Y., Gupta, B.B., Choi, C.: Visual saliency guided complex image retrieval. Pattern Recognit. Lett. **130**, 64–72 (2020)

112. Xu, G., Shi, Y., Sun, X., Shen, W.: Internet of things in marine environment monitoring: a review. Sensors **19**(7), 1711–1711 (2019)

113. Luhach, A.K., Hawari, K.B.G., Mihai, I.C., Hsiung, P.A., Mishra, R.B.: Smart Computational Strategies: Theoretical and Practical Aspects (2019)

114. Li, S., Da Xu, L., S.Z.: The internet of things: a survey. Inf. Syst. Front. **17**(2), 243–259 (2015). https://doi.org/10.1007/s10796-014-9492-7

115. Zhang, A.L.: Research on the architecture of internet of things applied in coal mine. In: 2016 International Conference on Information System and Artificial Intelligence (ISAI), pp. 21–23 (2016)

116. Sedik, A., Hammad, M., Abd El-Samie, F.E., Gupta, B.B., Abd El-Latif, A.A.: Efficient deep learning approach for augmented detection of coronavirus disease. Neural Comput. Appl. 1–18 (2021). https://doi.org/10.1007/s00521-020-05410-8

117. Stergiou, C.L., Psannis, K.E., Gupta, B.B.: IoT-based big data secure management in the fog over a 6G wireless network. IEEE Internet Things J. **8**(7), 5164–5171 (2021). https://doi.org/10.1109/jiot.2020.3033131

118. Atzori, L., Iera, A., Morabito, G.: The internet of things: a survey. Comput. Netw. **54**(15), 2787–2805 (2010). https://doi.org/10.1016/j.comnet.2010.05.010

119. Masud, M., Gaba, G.S., Alqahtani, S., Muhammad, G., Gupta, B.B., Kumar, P., Ghoneim, A.: A lightweight and robust secure key establishment protocol for internet of medical things in covid-19 patients care. IEEE Internet Things J. 1 (2021). https://doi.org/10.1109/jiot.2020.3047662

120. AlZu'bi, S., Shehab, M., Al-Ayyoub, M., Jararweh, Y., Gupta, B.: Parallel implementation for 3D medical volume fuzzy segmentation. Pattern Recognit. Lett. **130**, 312–318 (2020). https://doi.org/10.1016/j.patrec.2018.07.026

121. Gudivada, A., Philips, J., Tabrizi, N.: Developing concept enriched models for big data processing within the medical domain. Int. J. Softw. Sci. Comput. Intell. **12**(3), 55–71 (2020). https://doi.org/10.4018/ijssci.2020070105

122. John Sarivougioukas, A.V.: Modeling deep learning neural networks with denotational mathematics in ubihealth environment. Int. J. Softw. Sci. Comput. Intell. **12**(3), 14–27 (2020). https://doi.org/10.4018/ijssci.2020070102

123. Meriem Benadda, G.B.: Improving road safety for driver malaise and sleepiness behind the wheel using vehicular cloud computing and body area networks. Int. J. Softw. Sci. Comput. Intell. **12**(4), 19–41 (2020). https://doi.org/10.4018/ijssci.2020100102

124. Mohamed Sarrab, F.A.: Assisted-fog-based framework for IoT-based healthcare data preservation. Int. J. Cloud Appl. Comput. **11**(2), 1–16 (2021). https://doi.org/10.4018/ijcac.2021040101

125. Al-Shalabi, M.: COVID-19 symptoms monitoring mechanism using internet of things and wireless sensor networks. IJCSNS **20**(8), 16–16 (2020)

126. Maghded, H.S., Ghafoor, K.Z., Sadiq, A.S., Curran, K., Rawat, D.B., Rabie, K.: A novel AI-enabled framework to diagnose coronavirus COVID-19 using smartphone embedded sensors: design study. In: 2020 IEEE 21st International Conference on Information Reuse and Integration for Data Science (IRI) pp. 180–187 (2020)

127. Chowdhury, M.E., Rahman, T., Khandakar, A., Mazhar, R., Kadir, M.A., Mahbub, Z.B., Islam, K.R., Khan, M.S., Iqbal, A., Al Emadi, N. and Reaz, M.B.I.: Can AI help in screening viral and COVID-19 pneumonia? IEEE Access **8**, 132665–132676 (2020). https://doi.org/10.1109/access.2020.3010287

128. Ndiaye, M., Oyewobi, S.S., Abu-Mahfouz, A.M., Hancke, G.P., Kurien, A.M., Djouani, K.: IoT in the wake of COVID-19: a survey on contributions, challenges and evolution. IEEE Access **8**, 186821–186839 (2020). https://doi.org/10.1109/access.2020.3030090

129. Khan, R., Khan, S.U., Zaheer, R., Khan, S.: Future internet: the internet of things architecture, possible applications and key challenges. In: 2012 10th International Conference on Frontiers of Information Technology, pp. 257–260 (2012)

130. Dong, Y., Yao, Y.D.: IoT platform for COVID-19 prevention and control: a survey (2020)

131. Xia, F., Yang, L.T., L.W.A.V.: Internet of things. Int. J. Commun. Syst. **25**(9), 1101–1102 (2012). https://doi.org/10.1002/dac.2417

132. Abd El-Latif, A.A., Hossain, M.S. Wang, N.: Score level multibiometrics fusion approach for healthcare. Cluster Comput. **22**(1), 2425–2436 (2019)

133. Abd El-Latif, A.A., Abd-El-Atty, B., Venegas-Andraca, S.E., Elwahsh, H., Piran, M.J., Bashir, A.K., Song, O.Y., Mazurczyk, W.: Providing end-to-end security using quantum walks in IoT networks. IEEE Access **8**, 92687–92696 (2020). https://doi.org/10.1109/access.2020. 2992820

134. Citi, V., Martelli, A., Brancaleone, V., Brogi, S., Gojon, G., Montanaro, R., Morales, G., Testai, L., Calderone, V.: Anti-inflammatory and antiviral roles of hydrogen sulfide: rationale for considering H2S donors in COVID-19 therapy. Br J. Pharmacol. **177**(21), 4931–4941 (2020). https://doi.org/10.1111/bph.15230

135. Silva, A.F., Tavakoli, M.: Domiciliary hospitalization through wearable biomonitoring patches: recent advances, technical challenges, and the relation to COVID-19. Sensors **20**(23), 6835–6835 (2020). https://doi.org/10.3390/s20236835

136. Sandra Costanzo, A.F.: A non-contact integrated body-ambient temperature sensors platform to contrast COVID-19. Electronics **9**(10), 1658–1658 (2020). https://doi.org/10.3390/ electronics9101658

137. Ali, S., Singh, R.P., Javaid, M., Haleem, A., Pasricha, H., Suman, R., Karloopia, J.: A review of the role of smart wireless medical sensor network in COVID-19. J. Ind. Integr. Manag. **05**(04), 413–425 (2020). https://doi.org/10.1142/s2424862220300069

138. Ashton, K.: That 'internet of things' thing. RFID J. **22**(7), 97–114 (2009)

139. Cui, X.: The internet of things. In: Ethical Ripples of Creativity and Innovation, pp. 61–68 (2016)

140. Felix Wortmann, K.F.: Internet of things. Bus. Inf. Syst. Eng. **57**(3), 221–224 (2015). https:// doi.org/10.1007/s12599-015-0383-3

141. Anifah, L., Rusimamto, P.W., Nurhayati, S.I.H., Warju, H.: Internet of things-based monitoring system of patients using w1209 digital thermostat and pulse sensor. In: International Joint Conference on Science and Engineering, vol. 2020, 287–291 (2020)

142. Islam, M.M., Rahaman, A., Islam, M.R.: Development of smart healthcare monitoring system in IoT environment. SN Comput. Sci. **1**, 1–11 (2020)

143. Ghorbel, O., Ayedi, R., Chikha, H.B., Shehin, O., Frikha, M.: Design of a smart medical bracelet prototype for COVID-19 based on wireless sensor networks. Int. J. **9**(3), (2020)

144. Whitmore, A., Agarwal, A., Da Xu, L.: The internet of things—a survey of topics and trends. Inf. Syst. Front. **17**(2), 261–274 (2015). https://doi.org/10.1007/s10796-014-9489-2

145. Abd-El-Atty, B., Iliyasu, A.M., Alaskar, H., El-Latif, A., Ahmed, A.: A robust quasi-quantum walks-based steganography protocol for secure transmission of images on cloud-based e-healthcare platforms. Sensors **20**(11), 3108–3108 (2020). https://doi.org/10.3390/s20113108

146. Gershenfeld, N., Krikorian, R., Cohen, D.: The internet of things. Sci. Am. **291**(4), 76–81 (2004). https://doi.org/10.1038/scientificamerican1004-76

147. Coetzee, L., Eksteen, J.: The internet of things-promise for the future? an introduction. In: 2011 IST-Africa Conference Proceedings, pp. 1–9 (2011)

148. Greengard, S.: The Internet of Things (2015)

149. Kopetz, H.: Internet of things. Real-Time syst. 307–323 (2011)

150. Weber, R.H., R.W.: Internet of Things, vol. 12 (2010)

151. Holler, J., Tsiatsis, V.: Internet of Things, C.M.S.K.S.A.D.B. (2014)

152. Rose, K., Eldridge, S., Chapin, L.: The internet of things: an overview. The internet society (ISOC) **80**, 1–50 (2015)

153. Da Xu, L., He, W., Li, S.: Internet of things in industries: a survey. IEEE Trans. Ind. Inform. **10**(4), 2233–2243 (2014). https://doi.org/10.1109/tii.2014.2300753

154. Tan, L., Wang, N.: Future internet: the internet of things. In: 2010 3rd International Conference on Advanced Computer Theory and Engineering (ICACTE), vol. 5, pp. 5–376 (2010)

155. Mcewen, A., Cassimally, H.: Designing the Internet of Things (2013)

156. Chen, Y.K.: Challenges and opportunities of internet of things. In: 17th Asia and South Pacific Design Automation Conference, pp. 383–388 (2012)

157. Ray, P.P.: A survey on internet of things architectures. J. King Saud Univ.-Comput. Inf. Sci. **30**(3), 291–319 (2018)

158. Roman, R., Najera, P., Lopez, J.: Securing the internet of things. Computer **44**(9), 51–58 (2011). https://doi.org/10.1109/mc.2011.291
159. Stankovic, J.A.: Research directions for the internet of things. IEEE Internet Things J. **1**(1), 3–9 (2014). https://doi.org/10.1109/jiot.2014.2312291
160. Debasis Bandyopadhyay, J.S.: Internet of things: applications and challenges in technology and standardization. Wirel. Pers. Commun. **58**(1), 49–69 (2011). https://doi.org/10.1007/s11277-011-0288-5
161. Alaba, F.A., Othman, M., Hashem, I.A.T., Alotaibi, F.: Internet of things security: a survey. J. Netw. Comput. Appl. **88**, 10–28 (2017). https://doi.org/10.1016/j.jnca.2017.04.002
162. Fleisch, E.: What is the internet of things? an economic perspective. Econ. Manag. Financ. Mark. **5**(2), 125–157 (2010)
163. Sundmaeker, H., Guillemin, P., Friess, P., Woelffle, S.: Vision and challenges for realising the internet of things. cluster of European research projects on the internet of things. Euro. Comm. **3**(3), 34–36 (2010)
164. Haller, S.: The Things in the Internet of Things, pp. 26–30 (2005)
165. Manyika, J., Chui, M., Bisson, P., Woetzel, J., Dobbs, R., Bughin, J., Aharon, D.: Unlocking the potential of the internet of things (2015)
166. Morgan, J.: A Simple Explanation of the Internet of Things (2014)
167. Want, R., Schilit, B.N., Jenson, S.: Enabling the internet of things. Computer **48**(1), 28–35 (2015). https://doi.org/10.1109/mc.2015.12
168. Haller, S., Karnouskos, S., Schroth, C.: The internet of things in an enterprise context. In: Future Internet Symposium, pp. 14–28 (2008)
169. Gluhak, A., Krco, S., Nati, M., Pfisterer, D., Mitton, N., Razafindralambo, T.: A survey on facilities for experimental internet of things research. IEEE Commun. Mag. **49**(11), 58–67 (2011). https://doi.org/10.1109/mcom.2011.6069710
170. Perera, C., Liu, C.H., Jayawardena, S., Chen, M.: A survey on internet of things from industrial market perspective. IEEE Access **2**, 1660–1679 (2014). https://doi.org/10.1109/access.2015.2389854
171. Conti, J.P.: The internet of things. Commun. Eng. **4**(6), 20–25 (2006)
172. Gupta, D., Bhatt, S., Gupta, M., Tosun, A.S.: Future smart connected communities to fight covid-19 outbreak. Internet Things **13**, 100342–100342 (2021). https://doi.org/10.1016/j.iot.2020.100342
173. Ma, H.D.: Internet of things: objectives and scientific challenges. J. Comput. Sci. Technol. **26**(6), 919–924 (2011)
174. Shancang Li, Li Da Xu, S.Z.: 5g internet of things: A survey. Journal of Industrial Information Integration **10**, 1–9 (2018). https://doi.org/10.1016/j.jii.2018.01.005
175. Huang, Y., Li, G.: Descriptive models for internet of things. In: 2010 International Conference on Intelligent Control and Information Processing, pp. 483–486 (2010)
176. Anand Paul, R.J.: Internet of things: a primer. Hum. Behav. Emerg. Technol. **1**(1), 37–47 (2019). https://doi.org/10.1002/hbe2.133
177. Zanella, A., Bui, N., Castellani, A., Vangelista, L., Zorzi, M.: Internet of things for smart cities. IEEE Internet Things J. **1**(1), 22–32 (2014). https://doi.org/10.1109/jiot.2014.2306328
178. Schoenberger, C.R., B.U.: The internet of things. Forbes Mag. **169**(6), 155–160 (2002)
179. Gubbi, J., Buyya, R., Marusic, S., Palaniswami, M.: Internet of things (IoT): a vision, architectural elements, and future directions. Future Gener. Comput. Syst. **29**(7), 1645–1660 (2013). https://doi.org/10.1016/j.future.2013.01.010
180. Kranenburg, R.V.: The internet of things. World Affairs: J. Int. Issues **15**(4), 126–141 (2011)
181. Bunz, M., G.M.: The Internet of Things (2017)
182. Tsafack, N., Sankar, S., Abd-El-Atty, B., Kengne, J., Jithin, K.C., Belazi, A., Mehmood, I., Bashir, A.K., Song, O.Y., Abd El-Latif, A.A.: A new chaotic map with dynamic analysis and encryption application in internet of health things. IEEE Access **8**, 137731–137744 (2020). https://doi.org/10.1109/access.2020.3010794
183. Sun, Q.B., Liu, J., Li, S., Fan, C.X., Sun, J.J.: Internet of things: summarize on concepts, architecture and key technology problem. J. Beijing Univ. Posts Telecommun. **3**(3), 1–9 (2010)

184. Buyya, R., Dastjerdi, A.V.: Internet of Things: Principles and Paradigms (2016)
185. De, S., Barnaghi, P., Bauer, M., Meissner, S.: Service modelling for the internet of things. In: 2011 Federated Conference on Computer Science and Information Systems (FedCSIS), pp. 949–955 (2011)
186. Agrawal, S., Vieira, D.: A survey on internet of things. Abakós 1(2), 78–95 (2013). https://doi.org/10.5752/10.5752/p.2316-9451.2013v1n2p78
187. Yuehong, Y.I.N., Zeng, Y., Chen, X., Fan, Y.: The internet of things in healthcare: an overview. J. Ind. Inf. Integr. 1, 3–13 (2016)
188. Yang, D.L., Liu, F., Y.D.L.: A survey of the internet of things. In: Proceedings of the 1st International Conference on E-Business Intelligence (ICEBI2010) (2010)
189. Lee, G.M., Crespi, N., Choi, J.K., Boussard, M.: Internet of things. In: Evolution of Telecommunication Services, pp. 257–282 (2013)
190. Bari, N., Mani, G., Berkovich, S.: Internet of things as a methodological concept. In: 2013 Fourth International Conference on Computing for Geospatial Research and Application, pp. 48–55 (2013)
191. Lindqvist, U., Neumann, P.G: The future of the internet of things. Commun. ACM 60(2), 26–30 (2017). https://doi.org/10.1145/3029589
192. Waher, P.: Learning Internet of Things (2015)
193. Al-Fuqaha, A., Guizani, M., Mohammadi, M., Aledhari, M., Ayyash, M.: Internet of things: a survey on enabling technologies, protocols, and applications. IEEE Commun. Surv. Tutor. 17(4), 2347–2376 (2015). https://doi.org/10.1109/comst.2015.2444095
194. Singh, D., Tripathi, G., Jara, A.J..: A survey of internet-of-things: future vision, architecture, challenges and services. In: 2014 IEEE World Forum on Internet of Things (WF-IoT), pp. 287–292 (2014)
195. Uckelmann, D., Harrison, M., Michahelles, F.: Architecting the Internet of Things (2011)
196. Miorandi, D., Sicari, S., De Pellegrini, F., Chlamtac, I.: Internet of things: vision, applications and research challenges. Ad Hoc Netw. 10(7), 1497–1516 (2012). https://doi.org/10.1016/j.adhoc.2012.02.016
197. Al-Ayyoub, M., Al-Mnayyis, N., Alsmirat, M.A., Alawneh, K., Jararweh, Y., Gupta, B.B.: SIFT based ROI extraction for lumbar disk herniation cad system from MRI axial scans. J. Ambient Intell. Hum. Comput. 1–9 (2018). https://doi.org/10.1007/s12652-018-0750-2
198. Feki, M.A., Kawsar, F., Boussard, M., Trappeniers, L.: The internet of things: the next technological revolution. Computer 46(2), 24–25 (2013). https://doi.org/10.1109/mc.2013.63
199. Said, O., Masud, M.: Towards internet of things: survey and future vision. Int. J. Comput. Netw. 5(1), 1–17 (2013)
200. Islam, S.R., Kwak, D., Kabir, M.H., Hossain, M., Kwak, K.S.: The internet of things for health care: a comprehensive survey. IEEE Access 3, 678–708 (2015). https://doi.org/10.1109/access.2015.2437951
201. Baoyun, W.: Review on internet of things. J. Electron. Meas. Instrum. 12, 1–7 (2009)
202. Chaouchi, H.: The Internet of Things: connecting objects to the web (2013)
203. Chase, J.: The evolution of the internet of things. Texas Instrum. 1, 1–7 (2013)
204. Lee, I., Lee, K.: The internet of things (IoT): applications, investments, and challenges for enterprises 58, 431–440 (2015)
205. Nord, J.H., Koohang, A., Paliszkiewicz, J.: The internet of things: review and theoretical framework. Expert Syst. Appl. 133, 97–108 (2019). https://doi.org/10.1016/j.eswa.2019.05.014
206. Saarikko, T., Westergren, U.H., Blomquist, T.: The internet of things: are you ready for what's coming? Bus. Horiz. 60(5), 667–676 (2017). https://doi.org/10.1016/j.bushor.2017.05.010
207. Vermesan, O., Friess, P., Guillemin, P., Gusmeroli, S., Sundmaeker, H., Bassi, A., Jubert, I.S., Mazura, M., Harrison, M., Eisenhauer, M., et al.: Internet of things strategic research roadmap. Internet Things-Glob. Technol. Soc. Trends 1(2011), 9–52 (2011)
208. Li, X., Lu, R., Liang, X., Shen, X., Chen, J., Lin, X.: Smart community: an internet of things application. IEEE Commun. Mag. 49(11), 68–75 (2011)

209. Chen, X.Y., Jin, Z.G.: Research on key technology and applications for internet of things. Phys. Procedia **33**, 561–566 (2012)
210. Liu, Y., Zhou, G.: Key technologies and applications of internet of things. In: 2012 Fifth International Conference on Intelligent Computation Technology and Automation, pp. 197–200 (2012)
211. Ghoneim, A., Muhammad, G., Amin, S.U., Gupta, B.: Medical image forgery detection for smart healthcare. IEEE Commun. Mag. **56**(4), 33–37 (2018). https://doi.org/10.1109/mcom.2018.1700817
212. Kaushik, S., Gandhi, C.: Ensure hierarchal identity based data security in cloud environment. Int. J. Cloud Appl. Comput. (IJCAC) **9**(4), 21–36 (2019)
213. Al-Ayyoub, M., AlZu'bi, S., Jararweh, Y., Shehab, M.A., Gupta, B.B.: Accelerating 3D medical volume segmentation using GPUS. Multimed. Tools Appl. **77**(4), 4939–4958 (2018)
214. Kamarudin, M.H., Maple, C., Watson, T.: Hybrid feature selection technique for intrusion detection system. Int. J. High Perform. Comput. Netw. **13**(2), 232–240 (2019)
215. Goléa, N.E.H., Melkemi, K.E.: Roi-based fragile watermarking for medical image tamper detection. Int. J. High Perform. Comput. Netw. **13**(2), 199–210 (2019)
216. Wang, H., Li, Z., Li, Y., Gupta, B., Choi, C.: Visual saliency guided complex image retrieval. Pattern Recognit. Lett. **130**, 64–72 (2020)
217. Mani, N., Moh, M., Moh, T.S.: Defending deep learning models against adversarial attacks. Int. J. Softw. Sci. Comput. Intell. (IJSSCI) **13**(1), 72–89 (2021)
218. Abou-Nassar, E.M., Iliyasu, A.M., El-Kafrawy, P.M., Song, O.Y., Bashir, A.K., Abd El-Latif, A.A.: DITrust chain: towards blockchain-based trust models for sustainable healthcare IoT systems. IEEE Access **8**, 111223–111238 (2020)
219. Gudivada, A., Philips, J., Tabrizi, N.: Developing concept enriched models for big data processing within the medical domain. Int. J. Softw. Sci. Comput. Intell. (IJSSCI) **12**(3), 55–71 (2020)
220. Gou, Z., Yamaguchi, S., Gupta, B.: Analysis of various security issues and challenges in cloud computing environment: a survey. In: Identity Theft: breakthroughs in Research and Practice, pp. 221–247. IGI global (2017)
221. Zheng, Q., Wang, X., Khan, M.K., Zhang, W., Gupta, B.B., Guo, W.: A lightweight authenticated encryption scheme based on chaotic SCML for railway cloud service. IEEE Access **6**, 711–722 (2017)
222. Gupta, B., Quamara, M.: An overview of internet of things (IoT): architectural aspects, challenges, and protocols. Concurrency Comput.: Pract. Exp. **32**(21), e4946 (2020)
223. Divya, M., Subhash, N., Vishnu, P., Tejesh, P.: Smart health care monitoring based on internet of things (IoT). Int. J. Sci. Res. Eng. Dev. **3**(1), 409–414 (2020)
224. Bhuyan, M.H., Hasan, M.: Design and simulation of heartbeat measurement system using arduino microcontroller in proteus. Int. J. Biomed. Biol. Eng. **14**(10), 350–357 (2020)
225. Rahimoon, A.A., Abdullah, M.N., Taib, I.: Design of a contactless body temperature measurement system using arduino. Indones. J. Electr. Eng. Comput. Sci. **19**(3), 1251–1251 (2020). https://doi.org/10.11591/ijeecs.v19.i3.pp1251-1258
226. Abuzairi, T., Sumantri, N.I., Irfan, A., Mohamad, R.M.: Infrared thermometer on the wall (iThermowall): an open source and 3-D print infrared thermometer for fever screening pp. 168–168 (2021)
227. Hasan, D., Ismaeel, A.: Designing ECG monitoring healthcare system based on internet of things blynk application. J. Appl. Sci. Technol. Trends **1**(3), 106–111 (2020). https://doi.org/10.38094/jastt1336
228. Hadis, N.S.M., Amirnazarullah, M.N., Jafri, M.M., Abdullah, S.: IoT based patient monitoring system using sensors to detect, analyse and monitor two primary vital signs. J. Phys.: Conf. Ser. **1535**, 012004–012004 (2020). https://doi.org/10.1088/1742-6596/1535/1/012004
229. Sam, D., Srinidhi, S., Niveditha, V.R., Amudha, S., Usha, D.: Progressed IoT based remote health monitoring system. Int. J. Control Autom. **13**(2s), 268–273 (2020)
230. Mohammed, M.N., Syamsudin, H., Al-Zubaidi, S., AKS, R.R., Yusuf, E.: Novel COVID-19 detection and diagnosis system using iot based smart helmet. Int. J. Psychosoc. Rehabil. **24**(7), 2296–2303 (2020)

231. Mhatre, P., Shaikh, A., Khanvilkar, S.: Non Invasive E-Health Care Monitoring System Using IOT (2020)
232. Petrović, N., Kocić.: IoT-based system for COVID-19 indoor safety monitoring. preprint). IcETRAN **2020**, 1–6 (2020)

DDoS Attack Detection in Vehicular Ad-Hoc Network (VANET) for 5G Networks

Akshat Gaurav, B. B. Gupta, Francisco José García Peñalvo, Nadia Nedjah, and Konstantinos Psannis

Abstract VANET is a crucial part of the intelligent transport system (ITS). VANET helps the vehicle nodes to exchange important and life-saving information, so any attack on VANET should be detected fast. The DDoS attack is one of the cyber-attacks that attack the availability of the VANET systems. Due to the DDoS attack vehicle nodes are not capable to exchange valuable information. In this chapter, we propose a fog-based DDoS detection approach that uses fuzzy logic to differentiate attack traffic from normal traffic in 5G-enabled smart cities. The proposed approach achieves more than 90% precision and true negative rate, it indicates that our proposed approach correctly identifies the DDoS attack traffic.

Keywords VANET · 5G · Fog computing · DDoS · Fuzzy logic · Trust based system

1 Introduction

VANET [26, 29, 65] can be considered as a special class of MANET [18, 28] in which nodes are vehicles and they are moving on predefined roots. The foundation for

A. Gaurav
Ronin Institute, Montclair, USA
e-mail: akshat.gaurav@ronininstitute.org

B. B. Gupta (✉)
National Institute of Technology, Kurushatra, Haryana, India

F. J. G. Peñalvo
University of Salamanca, Salamanca, Spain
e-mail: fgarcia@usal.es

N. Nedjah
State University of Rio de Janeiro, Rio de Janeiro, Brazil
e-mail: nadia@eng.uerj.br

K. Psannis
University of Macedonia, Thessaloniki, Greece
e-mail: kpsannis@uom.gr

© The Author(s), under exclusive license to Springer Nature Switzerland AG 2022
A. A. Abd El-Latif et al. (eds.), *Security and Privacy Preserving for IoT and 5G Networks*,
Studies in Big Data 95, https://doi.org/10.1007/978-3-030-85428-7_11

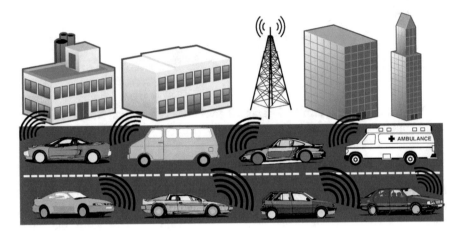

Fig. 1 VANET Architecture

the development of VANET was kept in 1999 when the US Federal Communication Commission allocated a 75 MHz bandwidth of 5.9 GHz band to Dedicated Short-Range Communication (DSRC). VANET attracted researcher's interest in 2001 when the standards committee of ASTM decided IEEE 802.11a as the working protocol for DSRC. In 2004, the IEEE amended the 802.11a protocol and started developing it as the Wireless Access in Vehicular Environments (WAVE) standard for VANET.

Providing safety to the passenger and providing better driving conditions to the driver are the key applications of VANET [12, 50]. VENT provides all essential information like nearby hospitals, petrol pump, road traffic jam information, to the driver of the vehicle due to which life and resources are saved. Due to the life-saving nature of VANET [32, 54], it attracts many researchers.

VANET consists of roadside stationary units (RSU) [57] and mobile vehicles that are equipped with on-board units (OBU) [42]. The vehicle can communicate through OBU with other neighboring nodes (V2V) or with the RSU (V2I) [41]. These communications are represented in the Fig. 1. The communication in VANET can be 'one-hop' i.e between two neighbor vehicles or 'multihop' in which vehicles re-transmits the received message. The RSU can also be used to increase the communication range or these units can also be used to connect the vehicle data to a cloud server [5, 60].

As VANET uses wireless communication, it is easily affected by different types of cyber-attacks . The DDoS attack is one of these types of attacks. DDoS attack affects V2V and V2I communication and due to this attack, the vehicle node is neither able to process the information received from its neighbors nor it can transfer important information to its neighbors. Therefore, due to the DDoS attack, the safety of the passengers is affected.

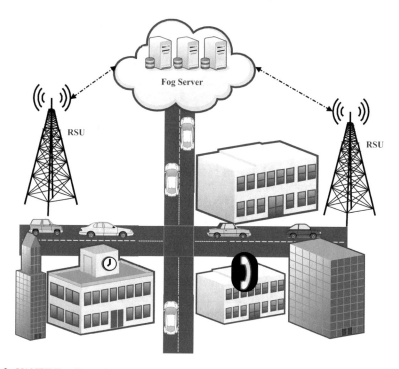

Fig. 2 VANET Fog Scenario

There are many ways to counter the DDoS attack [7, 9, 20, 33, 36, 66], one of them is fog computing [26]. Cisco [47] proposes a concept of decentralization computing known as fog computing, which brings storage, computing, and other cloud facilities near to the end device [14]. In fog computing, fog servers are installed near the end device, so that end devices latency reduces. Each fog server can be considered as a lightweight cloud server [30], because of these features, fog computing can be used for quick detection of a malicious node.

The concept of fog computing is also introduced in VANET. In VANET, fog nodes analyze the network and store the important information into fog servers as represented in Fig. 2. The fog servers which are near to the edge devices process the incoming information and quickly take a decision and detect the malicious vehicle node. With the development of 5G network [62] the use of fog devices increases, 5G network helps the fog devices in improving the quality of service and reducing the latency. Hence, fog devices can easily be used in 5G networks for implementing security measures.

In this paper, we propose a fog computing based DDoS [13, 27, 58] attack detection scheme for 5G networks. In our proposed approach, the fog server calculates the trust value for each node according to the fuzzy logic [8, 46]. The trust value of the vehicle node is used as the key factor for the identification of the DDoS attack traffic from normal traffic.

The rest of the paper is organized in the following sections, Sect. 2 represents the related work, Sect. 3 explains our proposed model in detail. Section 4 shows simulated results and comparison of our proposed work with other algorithms. Finally, Sect. 5 represents the conclusion.

2 Related Work

5G communication network [17, 48] provides many benefits like low latency, high bandwidth to IoT networks [2, 4, 11, 21], cloud computing [37], and fog computing [6, 44]. However, along with all benefits, the risks of cyber attacks are increased for 5G communication systems [24, 25, 34, 72]. Therefore, there is a need for security protocols that improve the security of devices in a 5G communication network [15, 35, 55, 61, 71]. In this section, we will review some security measures for VANET for 5G communication networks

Gu et al. [31] proposes a fog-based malicious node detection method in VANETs. In this approach, the cluster head transfers the vehicle node information and topology information to the fog server. At the fog server, the reputation value of all the nodes is calculated and according to this reputation value, a malicious node is identified.

Erskine et al. [23] proposes a fog-based DoS attack detection approach. In this approach, the author uses the Cuckoo/CSA Artificial Bee Colony (ABC) algorithm, Firefly/Genetic Algorithm (GA), and Feedforward back propagation neural network (FFBPNN) for the detection of DoS attack nodes in real-time.

Cui et al. [19] propose an efficient data downloading method in VANET by using a fog layer. In this scheme, the RSU finds the popular data by the request of the vehicle nodes and stores it at nearby edge devices, so if any vehicle node wants the data, then it can download it directly from the edge device. This will increase the downloading efficiency of the vehicle node.

Wang et al. [69] proposes a real-time traffic management scheme using fog in VANETs. In this scheme, the author considers the vehicle nodes near RSU as fog nodes, and both moving and parked nodes can be considered as fog nodes. If fog nodes are present, then RSU uploads the data from the vehicle nodes to the fog nodes due to which the processing time is reduced.

Khare et al. [38] proposes black-hole, wormhole, and DDoS detection methods in MEANETs. In this method, the author uses fuzzy logic to find the malicious node. In this method, the network is divided into different clusters, and nodes in each cluster have a trust value if any node starts behaving maliciously, then its neighboring nodes start decreasing its trust value according to the fuzzy logic. The node having the lowest trust value is isolated from the network.

Kolandaisamy et al. [42] proposed a cluster head controlled detection scheme for VANETs. In this scheme, the VANET network is divided into different clusters and each cluster has a cluster head. The cluster head creates the trace file for each node in the cluster. By using the trace file for each session, an Attack Signature Sample Rate (CCA) for each node is calculated and by using CCA, a malicious node is detected.

Rudraraju et al. [52] proposed a collaborative intrusion detection model for VANET. In this scheme, the author uses FireCol and DGSOT algorithm to detect a malicious node. FireCol analyzes the incoming traffic by a set of rules and changes the reputation value of vehicle nodes according to their behavior. The DGSOT algorithm uses a hierarchy-based approach for the detection of a malicious node.

Adhikary et al. [3] proposes a hybrid algorithm based on machine learning for the detection of a malicious node in VANET. The proposed approach is the combination of SVM kernel methods of AnovaDot and RBFDot. The proposed approach uses collision packet drop jitter for firstly, the training of the algorithm and then detection of the malicious vehicle node.

Singh et al. [59] proposes a novel algorithm EAPDA, for the identification and detection of the malicious node. In this scheme, RSU monitors the vehicle node around it and stores the node information like vehicle id, location, and timestamp in the database. During each session, RSU shares its database with the vehicle nodes in the database, then the vehicle nodes according to the database identifies the malicious nodes within the cluster.

RoselinMary et al. [51] proposes APDA, a novel method for the detection of a malicious node. In this method, RSU uses the APDA algorithm to monitor the nodes in the crystal. RSU prepares a database of each node in the cluster which stores the position and velocity of each vehicle node. Then RSU calculates the frequency of each vehicle node and then compares this frequency with a predefined threshold when the malicious node is detected.

Kumar et al. [43] proposes MMPDA, a novel method for the detection of a malicious node. In this method, the network is divided into a cluster head and verifier nodes. According to the MMPDA algorithm, if the bandwidth of the network is more than the predefined threshold value, then it is assumed that a DOS attack occurs. During the attack, the entropy of each node is calculated, and compared with the entropy value before the attack, if the new entropy value is less than the previous entropy value, then it is assumed that a vehicle node is under attack.

Liu et al. [45] provides a detailed analysis of secure federated learning in 5G communication systems. The paper gives the detailed analysis of the applications of federated learning in 5G networks. The proposed approach uses the concept of blockchain to provide security to the federated learning framework. Authors contributed to create a marketplace where users can solve the federated learning issues. Abd et al. [1] also points out the need for a security mechanism for 5G networks. The authors proposed a data encryption protocol that is based on quantum walks. The advantage of the proposed protocol is that through it the IoT devices can share the encrypted data using AKD protocol.

Khayyat et al. [39] draws attention to the point that there is a need of autonomous control for advance vehicle systems. To fulfills the need of autonomous vehicle system in 5G network, the authors proposed a multilevel cloud-based system. The proposed approach is based on the Tactile Internet and efficiently reduces the latency. Khayyat et al. [40] points out the issue of energy conservation and the optimal way of sharing resources. Authors also proposed an integration model of computational offloading for resource sharing. The proposed approach is is based on the deep learning approach [16, 49, 53, 56, 73, 74], which is different from the traditional offloading techniques [64, 68, 70, 76, 77]. Elgendy et al. [22] also proposed the computational offloading for mobile edge computing. The proposed technique uses the catching concept, in which the program is cached at the edge devices. Authors proposed Q-learning and deep-Q-network-based algorithms to solve the task offloading. Zhang et al. [75] proposed a security mechanism for task offloading. In the proposed approach, the author introduces a security layer to improve the security of task offloading.

3 Proposed Approach

3.1 Proposed Framework

In the proposed framework, each vehicle node transfers the data to its one-hop neighbor until the data reaches the fog nodes. In our proposed approach, RSU and parked vehicles are considered as the fog nodes. Finally, the fog nodes pass the information to the fog server and at the fog server processing of the information takes place. One fog server is connected to the fog nodes of multiple regions and all the fog servers are connected, so they can share information. Hence, if a malicious node moves from one region to another region, then also our proposed method can detect it. Figure 3 represents the framework of our proposed approach, fog server 1 and fog server 2 are connected to the fog nodes in regions 1 and 2 respectively. All fog nodes pass

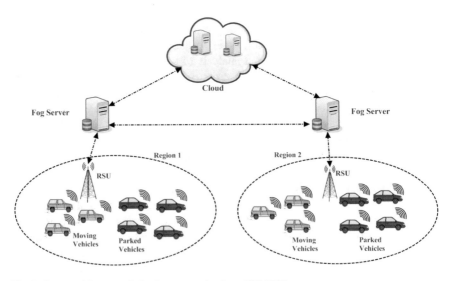

Fig. 3 Proposed framework for fog computing based VANET

the information to the fog servers and these fog servers also share the information among themselves, so if a vehicle node moves from one region to another region, then its information is not lost. This architecture helps to detect malicious node vehicles efficiently.

3.2 DDoS Attack Detection Approach

In this section, our proposed approach is explained. Our proposed approach is implemented on the fog servers, hence the efficiency of detection of the DDoS attack in the VANET network is increased. As explained in our proposed framework, fog servers are connected to the fog nodes of the region and fog servers of nearby regions, so implementing the proposed approach on the fog servers increases the detection range of the proposed approach, and it is easy to catch the malicious node vehicle. In our proposed approach, each incoming packet is analyzed in the following steps and Fig. 4 represents our proposed approach.

- Each node in the VANET has a trust value, this trust value is an integer that ranges between −0.5 and +2.
- When a new node is entered into the network, then the neighboring nodes assign a zero-trust value to the new node vehicle.
- Each node makes a database of the trust values of its neighboring node vehicles and shares this database with the fog node, which in turn shares it with the fog server. Therefore, finally, fog servers have the trust values of all nodes in the network.
- If a node behaves maliciously, then its neighboring node decreases its trust value according to the fuzzy logic. The updated trust value of the vehicle node is shared with the fog server.
- If the trust value of a node is at the lowest point, then it is considered as a malicious node and fog servers broadcast this information in all networks. Finally, the malicious node vehicle is blacklisted from the network.

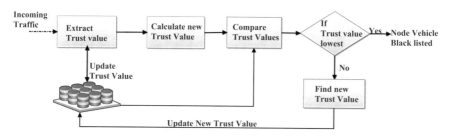

Fig. 4 Block diagram of proposed approach

3.3 Description of Algorithm

This section explains the algorithm that is used to detect the DDoS attack traffic. The attributes used in the algorithm are explained in the Table 1.

For each incoming packet (P_k) in the time window (T). Firstly, extract the IP address $(I P_k)$ of the source then find the counter value $(C_{I P_k})$ and trust value $(T_{I P_k})$ assigned to the source. If the source IP address is not present in the database, i.e., the source is new in the region, then set its trust value equal to the initial—trust value from the Table 2 and set the counter value assigned to it equals zero.

If the source IP address $(I P_k)$ is present in the database then with each incoming packet, the counter value $(C_{I P_k})$ assigned to the source is increased by one. Then use Algorithm 2 to determine the new trust value of the source. According to algorithm 2, if the counter value $(C_{I P_k})$ is less than the predefined threshold value (T) then the new trust value $(T_{I P_k})$ is increased by the trust update factor (δ) but if the counter value $(C_{I P_k})$ is more than the threshold value (T) then the new trust value is reduced by the trust update factor (δ). Algorithm 2 returns the new trust value, then algorithm 1 analyzes this new trust value $(T_{I P_k})$. If the new trust value $(T_{I P_k})$ is equal to untrusted, then the source of the packet is blacklisted and all packets coming from that ip address are dropped. The time complexity of the proposed approach is $\mathcal{O}(n)$, where n is the number of incoming packets in a specific time window.

Table 1 Parameters used in algorithm

Parameter	Value
P_k	kth *incomming packet*
P_{IP}	IP of kth *incoming packet*
C_{IP}	Counter
T_{IP}	Trust value
Th	Threshold value
δ	Trust update
F	Fuzzy set
T	Time window

Table 2 Fuzzy based trust assignment

Category	Crisp value
Un-trusted	−0.5
Initial-trust	0
Llow trust	0.5
Moderate trust	1.5
High trust	2

Algorithm 1: Algorithm to Analysis Incoming traffic
Input : Incoming Packets
Output : Weather IP address is blacklist or not
Start
for Every incoming packet (P_k) **do**
$\quad$ $P_k \rightarrow IP_k$, IP address is extracted
$\quad$ $P_k \rightarrow C_{IP_k}$, Counter value is extracted
$\quad$ $P_k \rightarrow T_{IP_k}$, Trust value is extracted
$\quad$ **if** C_{IP_k} *not defined* **then**
$\quad\quad$ Assign new counter C_{IP}
$\quad\quad$ $C_{IP_k} = 0$
$\quad\quad$ T_{IP_k} = initial–trust
$\quad$ **end**
$\quad$ **else**
$\quad\quad$ C_{IP_k} + +
$\quad\quad$ T_{IP_k} = Algorithm 2(C_{IP_k}, T_{IP_k})
$\quad$ **end**
$\quad$ **if** T_{IP_k} == *un-trusted* **then**
$\quad\quad$ IP address of kth source is blacklisted
$\quad\quad$ Filter out all the packets coming from the IP
$\quad\quad$ address
$\quad$ **end**
$\quad$ **else**
$\quad\quad$ IP address of kth source is trusted
$\quad$ **end**
end
END

Algorithm 2: Algorithm to find the Trust value
Input : Counter (C_{IP_k}) and trust value (T_{IP_k})
Output : Updated Trust value
Start
$F \rightarrow$ Fuzzy set
$C_{IP} \rightarrow C$, Counter value is extracted
$T_{IP_k} \rightarrow T_{old}$, Trust value is extracted
if $C < Th$ **then**
$\quad$ $T_{old} \rightarrow T_{old} + \delta$
$\quad$ **if** $T_{old} > trusted$ **then**
$\quad\quad$ return trusted
$\quad$ **end**
$\quad$ **else**
$\quad\quad$ return T_{old}
$\quad$ **end**
end
else
$\quad$ $T_{old} \rightarrow T_{old} - \delta$
$\quad$ **if** $T_{old} < un\text{-}trusted$ **then**
$\quad\quad$ return un-trusted
$\quad$ **end**
$\quad$ **else**
$\quad\quad$ return T_{old}
$\quad$ **end**
end
END

4 Results and Discussion

We used ONMET++ [67], Veins [63], and SUMO [10] for the simulation of our proposed approach. SUMO is a road traffic simulator which is used to simulate and analyse real world road traffic and other road management systems. Veins is an open source simulator which is used for vehicular network simulations. OMNET++ is an open source event-based simulator used to analyze discrete events and generate results. Veins is used to integrate OMNET++ and SUMO together and run the simulation. The simulation parameters are represented in Table 3. In the simulation, the attack packets try to flood the victim's vehicle with a large amount of fake traffic. In the simulation, the attacker node generates packets at every one second and, the legitimated nodes generate packets at every five seconds. The whole simulation is run for 200 seconds and all log values are saved. Our proposed approach is independent of the routing protocol, so we take a general routing protocol in the simulation.

Our proposed approach is evaluated by the following parameters.

- Average throughput—It is the average amount of data passed through a communication channel during a specific time slot. It is calculated by using Eq. 1 In our proposed approach, the throughput increases as the malicious node vehicles are blacklisted and then filtered out, as represented in Fig. 5.

$$\text{Throughput} = \frac{Nss \times Ps}{T} \tag{1}$$

Table 3 Simulation parameters

Term	Value
Simulation area	$250 \times 250 \; m^2$
Simulation time	200 s
Routing protocol	Random
Traffic generation	Random
Becon rate (normal node)	5 s
Becon rate (attacker)	1 s
MAC layer	802.11p
Network interface	OMNET++
Network mobility framework	Veins
Traffic generator	SUMO

where Nss = number of successful packets

Ps = average packet size

T = Time slot

- Packet Delivery Ratio (PDR)—It is the ratio of packets received at the destination to the packets sent from the source. Packet delivery ratio is calculated by Eq. 2. In our proposed approach, with every time slot, malicious nodes are blacklisted so the number of packets sent by legitimated nodes increases hence PDR increases with each time slot. This is represented in Fig. 5.

$$PDR = \frac{Total\ Packet\ received}{Total\ Packet\ sent} \tag{2}$$

- Precision—It measures the accuracy of the algorithm. It is calculated by the Eq. 3. In our proposed approach the true positive is increases with each time slot and false positive is decreasing with each time slot, hence the precision value is increasing with time as represented in Fig. 5.

$$Precision = \frac{True\ Positive}{True\ Positive + False\ Positive} \tag{3}$$

- True Negative Rate (TNR)—It is the ratio of attack packets discarded by the node to the attack packets received by the node. The value of TNR is calculated by Eq. 4. In our proposed approach detection rate is increases with every time slot so the number of discarded attack packets increases with time. Hence, as shown in Fig. 5 TNR value is increased with time.

$$TNR = \frac{Attack\ packets\ discarded}{Total\ attack\ packet\ recived} \tag{4}$$

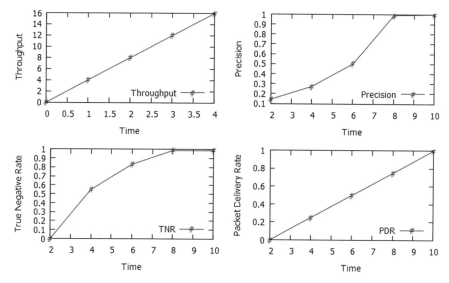

Fig. 5 Performance analyse of proposed approach

4.1 Result Comparison

In Table 4, our proposed work is compared with the existing techniques. The benefits of our proposed approach are as follows:

- Fast response—As our proposed approach is not using the clustering method, so it is less time-consuming compared to the other techniques which used clustering [31, 42, 43].
- Less complies—As our approach used a simple entropy-based response, so it is less complex and does not overload the vehicle node, unlike the techniques which use cryptographic [19] and machine learning [3, 23]
- Distributed approach-Our proposed approach uses a distributed technique and not requires any third-party control [52], which makes it more effective in a real-world scenario.

5 Conclusion

In VANET, the presence of a malicious node can affect the efficiency of the vehicle and the safety of the passengers in the vehicle. A DDoS attack is one of the attacks that is used by the attacker to attack the availability of the victim's vehicle by consuming its available resources. In this paper, we proposed a fog-based DDoS detection method in VANET for 5G networks. In our proposed approach, the fog server uses

Table 4 Result comparison

Properties Authors	Gu et al. [31]	Erskine and Elleithy [23]	Cui et al. [19]	Khare et al. [38]	Kolandaisamy et al. [42]	Rudraraju [52]	Adhikary et al. [3]	Kumar and Mann [43]	Our paper
Clustering overheads	✓	✗	✗	✗	✓	✗	✗	✓	✗
Statical methods	✗	✗	✗	✗	✗	✗	✗	✓	✓
Third party control	✗	✗	✗	✗	✗	✓	✗	✗	✗
Fuzzy logic	✗	✗	✓	✓	✗	✗	✗	✗	✓
Flooding attack detection	✗	✓	✗	✓	✓	✓	✓	✓	✓
Training dataset required	✗	✓	✗	✗	✗	✗	✓	✗	✗
Cryptographic method	✗	✗	✓	✗	✗	✗	✗	✗	✗

a trust-based system to differentiate legitimate nodes from the malicious nodes. Fog servers give the vehicle nodes a trust value which depends upon the node's behavior in the network, if the trust value is very low, then that node vehicle is considered as a malicious node. In our proposed approach, fog servers pass the information about malicious nodes to other fog servers, hence the malicious node is blacklisted from all regions of the VANET. We evaluated the proposed approach using throughput, packet delivery rate, packet drop ratio, precision, and false-positive rate. High precision and true negative rate rate indicate that our proposed approach efficiently detects the malicious node.

References

1. Abd EL-Latif, A.A., Abd-El-Atty, B., Venegas-Andraca, S.E., Mazurczyk, W.: Efficient quantum-based security protocols for information sharing and data protection in 5G networks. Future Gener. Comput. Syst. **100**, 893–906 (2019)
2. Abou-Nassar, E.M., Iliyasu, A.M., El-Kafrawy, P.M., Song, O.Y., Bashir, A.K., Abd El-Latif, A.A.: Ditrust chain: towards blockchain-based trust models for sustainable healthcare IoT systems. IEEE Access **8**, 111223–111238 (2020)
3. Adhikary, K., Bhushan, S., Kumar, S., Dutta, K.: Hybrid algorithm to detect DDoS attacks in vanets. Wirel. Person. Commun. 1–22 (2020)
4. Ahmed, S.H., Rani, S.: A hybrid approach, smart street use case and future aspects for internet of things in smart cities. Future Gener. Comput. Syst. **79**, 941–951 (2018)
5. Al-Nawasrah, A., Almomani, A.A., Atawneh, S., Alauthman, M.: A survey of fast flux botnet detection with fast flux cloud computing. Int. J. Cloud Appl. Comput. (IJCAC) **10**(3), 17–53 (2020)
6. Al-Turjman, F.: 5G-enabled devices and smart-spaces in social-IoT: an overview. Future Gener. Comput. Syst. **92**, 732–744 (2019)
7. Alomari, E., Manickam, S., Gupta, B., Karuppayah, S., Alfaris, R.: Botnet-based distributed denial of service (DDoS) attacks on web servers: classification and art (2012). arXiv preprint arXiv:1208.0403
8. AlZu'bi, S., Shehab, M., Al-Ayyoub, M., Jararweh, Y., Gupta, B.: Parallel implementation for 3D medical volume fuzzy segmentation. Pattern Recogn. Lett. **130**, 312–318 (2020)
9. Badve, O.P., Gupta, B.: Taxonomy of recent DDoS attack prevention, detection, and response schemes in cloud environment. In: Proceedings of the International Conference on Recent Cognizance in Wireless Communication & Image Processing, pp. 683–693. Springer (2016)
10. Behrisch, M., Bieker, L., Erdmann, J., Krajzewicz, D.: Sumo—simulation of urban mobility: an overview. In: Proceedings of SIMUL 2011, The Third International Conference on Advances in System Simulation. ThinkMind (2011)
11. Bello, O., Zeadally, S.: Toward efficient smartification of the internet of things (IoT) services. Future Gener. Comput. Syst. **92**, 663–673 (2019)
12. Benadda, M., Belalem, G.: Improving road safety for driver malaise and sleepiness behind the wheel using vehicular cloud computing and body area networks. Int. J. Softw. Sci. Comput. Intell. (IJSSCI) **12**(4), 19–41 (2020)
13. Bhushan, K., Gupta, B.B.: Distributed denial of service (DDoS) attack mitigation in software defined network (SDN)-based cloud computing environment. J. Ambient Intell. Hum. Comput. **10**(5), 1985–1997 (2019)
14. Bonomi, F., Milito, R., Zhu, J., Addepalli, S.: Fog computing and its role in the internet of things. In: Proceedings of the First Edition of the MCC Workshop on Mobile Cloud Computing. pp. 13–16 (2012)

15. Chaudhary, R., Kumar, N., Zeadally, S.: Network service chaining in fog and cloud computing for the 5G environment: data management and security challenges. IEEE Commun. Mag. **55**(11), 114–122 (2017)
16. Chen, J., Ran, X.: Deep learning with edge computing: a review. Proc. IEEE **107**(8), 1655–1674 (2019)
17. Chettri, L., Bera, R.: A comprehensive survey on internet of things (IoT) toward 5D wireless systems. IEEE Internet Things J. **7**(1), 16–32 (2019)
18. Chhabra, M., Gupta, B., Almomani, A.: A novel solution to handle DDoS attack in manet (2013)
19. Cui, J., Wei, L., Zhong, H., Zhang, J., Xu, Y., Liu, L.: Edge computing in vanets-an efficient and privacy-preserving cooperative downloading scheme. IEEE J. Sel. Areas Commun. **38**(6), 1191–1204 (2020)
20. Dahiya, A., Gupta, B.: Multi attribute auction based incentivized solution against DDoS attacks. Comput. Secur. **92**, 101763 (2020)
21. Dao, N.N., Park, M., Kim, J., Paek, J., Cho, S.: Resource-aware relay selection for inter-cell interference avoidance in 5G heterogeneous network for internet of things systems. Future Gener. Comput. Syst. **93**, 877–887 (2019)
22. Elgendy, I.A., Zhang, W.Z., He, H., Gupta, B.B., Abd El-Latif, A.A.: Joint computation offloading and task caching for multi-user and multi-task MEC systems: reinforcement learning-based algorithms. Wirel. Netw. 1–16 (2021)
23. Erskine, S.K., Elleithy, K.M.: Secure intelligent vehicular network using fog computing. Electronics **8**(4), 455 (2019)
24. Fang, D., Qian, Y., Hu, R.Q.: Security for 5G mobile wireless networks. IEEE Access **6**, 4850–4874 (2017)
25. Ferrag, M.A., Maglaras, L., Argyriou, A., Kosmanos, D., Janicke, H.: Security for 4G and 5G cellular networks: a survey of existing authentication and privacy-preserving schemes. J. Netw. Comput. Appl. **101**, 55–82 (2018)
26. Gaurav, A., Gupta, B., Castiglione, A., Psannis, K., Choi, C.: A novel approach for fake news detection in vehicular ad-hoc network (vanet). In: International Conference on Computational Data and Social Networks. pp. 386–397. Springer (2020)
27. Gaurav, A., Singh, A.K.: Super-router: A collaborative filtering technique against DDoS attacks. In: International Conference on Advanced Informatics for Computing Research. pp. 294–305. Springer (2017)
28. Gaurav, A., Singh, A.K.: Light weight approach for secure backbone construction for manets. J. King Saud Univ. Comput. Inf. Sci. (2018)
29. Ghori, M.R., Zamli, K.Z., Quosthoni, N., Hisyam, M., Montaser, M.: Vehicular ad-hoc network (vanet). In: 2018 IEEE International Conference on Innovative Research and Development (ICIRD). pp. 1–6. IEEE (2018)
30. Gou, Z., Yamaguchi, S., Gupta, B.: Analysis of various security issues and challenges in cloud computing environment: a survey. In: Identity Theft: breakthroughs in Research and Practice, pp. 221–247. IGI global (2017)
31. Gu, K., Dong, X., Jia, W.: Malicious node detection scheme based on correlation of data and network topology in fog computing-based vanets. IEEE Trans. Cloud Comput. (2020)
32. Gudivada, A., Philips, J., Tabrizi, N.: Developing concept enriched models for big data processing within the medical domain. Int. J. Softw. Sci. Comput. Intell. (IJSSCI) **12**(3), 55–71 (2020)
33. Gupta, B.B., Joshi, R.C., Misra, M.: An efficient analytical solution to thwart DDoS attacks in public domain. In: Proceedings of the International Conference on Advances in Computing, Communication and Control. pp. 503–509 (2009)
34. Ji, X., Huang, K., Jin, L., Tang, H., Liu, C., Zhong, Z., You, W., Xu, X., Zhao, H., Wu, J., et al.: Overview of 5G Secur. Technolo. Sci. China Inf. Sci. **61**(8), 1–25 (2018)
35. Jover, R.P.: The current state of affairs in 5g security and the main remaining security challenges (2019). arXiv preprint arXiv:1904.08394

36. Kamarudin, M.H., Maple, C., Watson, T.: Hybrid feature selection technique for intrusion detection system. Int. J. High Perform. Comput. Netw. **13**(2), 232–240 (2019)
37. Kaushik, S., Gandhi, C.: Ensure hierarchal identity based data security in cloud environment. Int. J. Cloud Appl. Comput. (IJCAC) **9**(4), 21–36 (2019)
38. Khare, A.K., Rana, J., Jain, R.: Detection of wormhole, blackhole and ddos attack in manet using trust estimation under fuzzy logic methodology. Int. J. Comput. Netw. Inf. Secur. **9**(7), 29 (2017)
39. Khayyat, M., Alshahrani, A., Alharbi, S., Elgendy, I., Paramonov, A., Koucheryavy, A.: Multi-level service-provisioning-based autonomous vehicle applications. Sustainability **12**(6), 2497 (2020)
40. Khayyat, M., Elgendy, I.A., Muthanna, A., Alshahrani, A.S., Alharbi, S., Koucheryavy, A.: Advanced deep learning-based computational offloading for multilevel vehicular edge-cloud computing networks. IEEE Access **8**, 137052–137062 (2020)
41. Kolandaisamy, R., Md Noor, R., Ahmedy, I., Ahmad, I., Reza Z'aba, M., Imran, M., Alnuem, M.: A multivariant stream analysis approach to detect and mitigate DDoS attacks in vehicular ad hoc networks. Wirel. Commun. Mob. Comput. **2018** (2018)
42. Kolandaisamy, R., Noor, R.M., Kolandaisamy, I., Ahmedy, I., Kiah, M.L.M., Tamil, M.E.M., Nandy, T.: A stream position performance analysis model based on ddos attack detection for cluster-based routing in vanet. J. Ambient Intell. Hum. Comput. 1–14 (2020)
43. Kumar, S., Mann, K.S.: Detection of multiple malicious nodes using entropy for mitigating the effect of denial of service attack in vanets. In: 2018 4th International Conference on Computing Sciences (ICCS), pp. 72–79. IEEE (2018)
44. Li, S., Da Xu, L., Zhao, S.: 5G internet of things: a survey. J. Ind. Inf. Integr. **10**, 1–9 (2018)
45. Liu, Y., Peng, J., Kang, J., Iliyasu, A.M., Niyato, D., Abd El-Latif, A.A.: A secure federated learning framework for 5G networks. IEEE Wirel. Commun. **27**(4), 24–31 (2020)
46. Mani, N., Moh, M., Moh, T.S.: Defending deep learning models against adversarial attacks. Int. J. Softw. Sci. Comput. Intell. (IJSSCI) **13**(1), 72–89 (2021)
47. Mukherjee, M., Matam, R., Shu, L., Maglaras, L., Ferrag, M.A., Choudhury, N., Kumar, V.: Security and privacy in fog computing: challenges. IEEE Access **5**, 19293–19304 (2017)
48. Nkenyereye, L., Liu, C.H., Song, J.: Towards secure and privacy preserving collision avoidance system in 5G fog based internet of vehicles. Future Gener. Comput. Syste. **95**, 488–499 (2019)
49. Peng, H., Shen, X.S.: Deep reinforcement learning based resource management for multi-access edge computing in vehicular networks. IEEE Trans. Netw. Sci. Eng. (2020)
50. Ponikwar, C., Hof, H.J.: Overview on security approaches in intelligent transportation systems (2015). arXiv preprint arXiv:1509.01552
51. RoselinMary, S., Maheshwari, M., Thamaraiselvan, M.: Early detection of DOS attacks in vanet using attacked packet detection algorithm (APDA). In: 2013 International Conference on Information Communication and Embedded Systems (ICICES). pp. 237–240. IEEE (2013)
52. Rudraraju, C.: Simulation of Detecting and Preventing DDoS in Vehicular Ad-hoc Networks (VANETS). Ph.D. thesis, Dublin, National College of Ireland (2020)
53. Sarivougioukas, J., Vagelatos, A.: Modeling deep learning neural networks with denotational mathematics in ubihealth environment. Int. J. Softw. Sci. Comput. Intell. (IJSSCI) **12**(3), 14–27 (2020)
54. Sarrab, M., Alshohoumi, F.: Assisted-fog-based framework for IoT-based healthcare data preservation. Int. J. Cloud Appl. Comput. (IJCAC) **11**(2), 1–16 (2021)
55. Schinianakis, D.: Alternative security options in the 5G and IoT era. IEEE Circuits Syst. Mag. **17**(4), 6–28 (2017)
56. Sedik, A., Hammad, M., Abd El-Samie, F.E., Gupta, B.B., Abd El-Latif, A.A.: Efficient deep learning approach for augmented detection of coronavirus disease. Neural Comput. Appl. 1–18 (2021)
57. Shakshuki, E.M., Kang, N., Sheltami, T.R.: EAACK—a secure intrusion-detection system for MANETs. IEEE Trans. Ind. Electron. **60**(3), 1089–1098 (2012)
58. Shidaganti, G.I., Inamdar, A.S., Rai, S.V., Rajeev, A.M.: SCEF: a model for prevention of DDOS attacks from the cloud. Int. J. Cloud Appl. Comput. (IJCAC) **10**(3), 67–80 (2020)

59. Singh, A., Sharma, P.: A novel mechanism for detecting dos attack in vanet using enhanced attacked packet detection algorithm (eapda). In: 2015 2nd International Conference on Recent Advances in Engineering Computational Sciences (RAECS), pp. 1–5 (2015)
60. Singh, A., Kumar, R.: A two-phase load balancing algorithm for cloud environment. Int. J. Softw. Sci. Comput. Intell. (IJSSCI) **13**(1), 38–55 (2021)
61. Sk, A., Masilamani, V.: A novel digital watermarking scheme for data authentication and copyright protection in 5G networks. Comput. Electr. Eng. **72**, 589–605 (2018)
62. Sodhro, A.H., Pirbhulal, S., Sangaiah, A.K., Lohano, S., Sodhro, G.H., Luo, Z.: 5G-based transmission power control mechanism in fog computing for internet of things devices. Sustainability **10**(4), 1258 (2018)
63. Sommer, C., German, R., Dressler, F.: Bidirectionally coupled network and road traffic simulation for improved IVC analysis. IEEE Trans. Mob. Comput. **10**(1), 3–15 (2010)
64. Sun, J., Gu, Q., Zheng, T., Dong, P., Valera, A., Qin, Y.: Joint optimization of computation offloading and task scheduling in vehicular edge computing networks. IEEE Access **8**, 10466–10477 (2020)
65. Tanwar, S., Vora, J., Tyagi, S., Kumar, N., Obaidat, M.S.: A systematic review on security issues in vehicular ad hoc network. Secur. Priv. **1**(5), e39 (2018)
66. Tupakula, U., Varadharajan, V., Mishra, P.: Securing SDN controller and switches from attacks. Int. J. High Perform. Comput. Netw. **14**(1), 77–91 (2019)
67. Varga, A.: A practical introduction to the omnet++ simulation framework. In: Recent Advances in Network Simulation, pp. 3–51. Springer (2019)
68. Wang, J., Feng, D., Zhang, S., Tang, J., Quek, T.Q.: Computation offloading for mobile edge computing enabled vehicular networks. IEEE Access **7**, 62624–62632 (2019)
69. Wang, X., Ning, Z., Wang, L.: Offloading in internet of vehicles: a fog-enabled real-time traffic management system. IEEE Trans. Ind. Inform. **14**(10), 4568–4578 (2018)
70. Xu, X., Xue, Y., Qi, L., Yuan, Y., Zhang, X., Umer, T., Wan, S.: An edge computing-enabled computation offloading method with privacy preservation for internet of connected vehicles. Future Gener. Comput. Syst. **96**, 89–100 (2019)
71. Yan, Z., Xie, H., Zhang, P., Gupta, B.B.: Flexible data access control in D2D communications. Future Gener. Comput. Syst. **82**, 738–751 (2018)
72. Yang, S., Yin, D., Song, X., Dong, X., Manogaran, G., Mastorakis, G., Mavromoustakis, C.X., Batalla, J.M.: Security situation assessment for massive MIMO systems for 5G communications. Future Gener. Comput. Syst. **98**, 25–34 (2019)
73. Ye, H., Li, G.Y., Juang, B.H.F.: Deep reinforcement learning based resource allocation for V2V communications. IEEE Trans. Veh. Technol. **68**(4), 3163–3173 (2019)
74. Zhan, W., Luo, C., Wang, J., Wang, C., Min, G., Duan, H., Zhu, Q.: Deep-reinforcement-learning-based offloading scheduling for vehicular edge computing. IEEE Internet Things J. **7**(6), 5449–5465 (2020)
75. Zhang, W.Z., Elgendy, I.A., Hammad, M., Iliyasu, A.M., Du, X., Guizani, M., Abd El-Latif, A.A.: Secure and optimized load balancing for multi-tier IoT and edge-cloud computing systems. IEEE Internet Things J. (2020)
76. Zhao, J., Li, Q., Gong, Y., Zhang, K.: Computation offloading and resource allocation for cloud assisted mobile edge computing in vehicular networks. IEEE Trans. Veh. Technol. **68**(8), 7944–7956 (2019)
77. Zhou, H., Chen, X., He, S., Chen, J., Wu, J.: DRAIM: a novel delay-constraint and reverse auction-based incentive mechanism for WiFi offloading. IEEE J. Sel. Areas Commun. **38**(4), 711–722 (2020)

Printed in the United States
by Baker & Taylor Publisher Services